SpringerBriefs in Mathematics

Series editors

Nicola Bellomo, Torino, Italy
Michele Benzi, Atlanta, USA
Palle Jorgensen, Iowa City, USA
Tatsien Li, Shanghai, China
Roderick Melnik, Waterloo, Canada
Lothar Reichel, Kent, USA
Otmar Scherzer, Vienna, Austria
Benjamin Steinberg, New York, USA
Yuri Tschinkel, New York, USA
George Yin, Detroit, USA
Ping Zhang, Kalamazoo, USA

SpringerBriefs in Mathematics showcases expositions in all areas of mathematics and applied mathematics. Manuscripts presenting new results or a single new result in a classical field, new field, or an emerging topic, applications, or bridges between new results and already published works, are encouraged. The series is intended for mathematicians and applied mathematicians.

BCAM SpringerBriefs

BCAM *SpringerBriefs* aims to publish contributions in the following disciplines: Applied Mathematics, Finance, Statistics and Computer Science. BCAM has appointed an Editorial Board, who evaluate and review proposals.

Typical topics include: a timely report of state-of-the-art analytical techniques, bridge between new research results published in journal articles and a contextual literature review, a snapshot of a hot or emerging topic, a presentation of core concepts that students must understand in order to make independent contributions.

Please submit your proposal to the Editorial Board or to Francesca Bonadei, Executive Editor Mathematics, Statistics, and Engineering: francesca.bonadei@springer.com

basque center for applied **mathematics**

More information about this series at http://www.springer.com/series/10030

Bernard Bonnard · Monique Chyba
Jérémy Rouot

Geometric and Numerical Optimal Control

Application to Swimming at Low Reynolds
Number and Magnetic Resonance Imaging

basque center for applied mathematics

Springer

Bernard Bonnard
Institut de Mathématiques de Bourgogne
Université de Bourgogne Franche-Comté
Dijon, France

Jérémy Rouot
Department of Applied Mathematics
EPF Graduate School of Engineering
Rosières-près-Troyes, France

Monique Chyba
Department of Mathematics
University of Hawaii at Manoa
Honolulu, HI, USA

ISSN 2191-8198   ISSN 2191-8201   (electronic)
SpringerBriefs in Mathematics
ISBN 978-3-319-94790-7   ISBN 978-3-319-94791-4   (eBook)
https://doi.org/10.1007/978-3-319-94791-4

Library of Congress Control Number: 2018948130

This Springer imprint is published by the registered company Springer Nature Switzerland AG
The registered company address is: Gewerbestrasse 11, 6330 Cham, Switzerland

Preface

The motivation for the notes presented in this volume of BCAM SpringerBriefs comes from a multidisciplinary graduate course offered to students in Mathematics, Physics or Control Engineering (at the University of Burgundy, France, and at the Institute of Mathematics for Industry, Fukuoka, Japan). The content is based on two real applications, which are the subject of current academic research programs and are motivated by industrial uses. The objective of these notes is to introduce the reader to techniques of geometric optimal control as well as to provide an exposure to the applicability of numerical schemes implemented in HamPath [32], Bocop [19], and GloptiPoly [47] software.

To highlight the main ideas and concepts, the presentation is restricted to the fundamental techniques and results. Moreover, the selected applications drive the exposition of the different methodologies. They have received significant attention recently and are promising, paving the way for further research by our potential readers. The applications have been chosen based on the existence of accurate mathematical models to describe them, models that are suitable for a geometric analysis, and the possibility of implementing results from the analysis in a practical manner.

The notes are self-contained; moreover, the simpler geometric computations can be reproduced by the reader using our presentation of the maximum principle. The weak maximum principle covers the case of an open control domain which is the standard situation encountered in the classical calculus of variations, and is suitable for analysis of the first application, motility at low Reynolds number, although a good understanding of the so-called transversality conditions is necessary. For the second application, control of the spin dynamics by magnetic fields in nuclear magnetic resonance, the use of the general maximum principle is required since the control domain is bounded. At a more advanced level, the reader has to be familiar with the numerical techniques implemented in the software used for our calculations. In addition, symbolic methods have to be used to handle the more complex computations.

The first application is the swimming problem at low Reynolds number describing the swimming techniques of microorganisms. It can be easily observed in nature, but also mechanically reproduced using robotic devices, and it is linked to medical applications. This example serves as an introduction to geometric optimal control applied to *sub-Riemannian geometry*, a non-trivial extension of Riemannian geometry and a 1980s tribute of control theory to geometry under the influence of R. Brockett [31]. We consider the *Purcell swimmer* [78], a three-link model where the shape variables are the two links at the extremities and the displacement is modeled by both the position and the orientation of the central link representing the body of the swimmer. To make a more complete analysis in the framework of geometric control, we use a simplified model from D. Takagi called the *copepod swimmer* [87], where only line displacement is considered using symmetric shape links, and which is also the swimming model for an abundant variety of zoo-plankton (copepods). This fact is particularly interesting with respect to validating the correlation between the *observed* and *predicted* displacements using the mathematical model. Also from the mathematical point of view, the copepod model is the *simplest slender body model* and can be obtained as a limit case of more complex systems, e.g., the Purcell swimmer.

For this problem, only the *weak maximum principle* is necessary and thus will be presented first, with its simple proof nevertheless containing all the geometric ingredients of the general maximum principle (see the historical paper by R.V. Gamkrelidze about the discovery of the maximum principle [40]). Moreover, in this case, under proper regularity assumptions, the second-order conditions can be easily explained and numerically implemented using the concepts of conjugate points and the Jacobi equation. More specifically, using the optimal control framework, the sub-Riemannian problem is expressed as:

$$\frac{dx}{dt}(t) = \sum_{i=1,\cdots,m} u_i(t)F_i(x(t)), \qquad \min_{u(.)} \int_0^T \sum_{i=1,\cdots,m} u_i^2(t)dt,$$

where $x \in M$, M is a smooth manifold, and the sub-Riemannian metric is defined by the orthonormal sub-frame $\{F_1, \cdots, F_m\}$ that determines the so-called non-holonomic constraints on the set of curves: $\dot{x}(t) \in D(x(t))$, where D is the distribution span$\{F_1, \cdots, F_m\}$. The relation to the swimming problem, modeled by some of the earliest prominent scientists (e.g., Newton, Euler, and more recently Stokes, Reynolds, and Purcell), is straightforward in the framework of control theory. The state variable x of the system decomposes into (x_1, x_2) where x_1 represents the displacement of the swimmer and x_2 is the shape variable representing the periodic shape deformation of the swimmer's body (called stroke) necessary to produce a net displacement for a given stroke. The mathematical model relates the speed of the displacement of x_1 to the speed of the shape deformation x_2, thus characterizing the sub-Riemannian constraints, while the expended mechanical energy defines the metric. The model comes from hydrodynamics and is subject to vital approximations. First, at the scale of the micro-swimmer's life, it implies that inertia can be neglected [45]. Second, according to the resistive force theory [44], the interaction of the swimmer with the fluid is reduced to a drag force depending linearly upon the velocity. Finally, each of our swimmers is approximated by a slender body. Due to these approximations, experiments are crucial to validate the models. This theoretical research also prompted experimentation using mechanical prototype devices (see for instance [75]).

Our objective is to provide a self-contained presentation of the model, of the underlying concepts of sub-Riemannian geometry, and of the techniques needed to conduct the mathematical analysis. The application of optimization techniques to the problem is recent. Our contribution's goal is to present a complete analysis using geometric and numerical techniques in the case of the copepod swimmer. It provides an excellent introduction to these methods, which have to be developed in the case of the Purcell swimmer based on our numerical results.

The second example concerns the optimal control of systems in *nuclear magnetic resonance* (NMR) and *magnetic resonance imaging* (MRI). The dynamics is modeled using the *Bloch equation* (1946), which describes at the macroscopic scale the evolution of the magnetization vector of a spin 1/2 particle depending on two relaxation parameters T_1 and T_2, which are the chemical signatures of the chemical species (e.g., water, fat) and controlled by an RF-magnetic pulse perpendicular to the strong polarizing field applied in the z−axis direction (see Bloch equations [18]). At the experimental level, optimal control was introduced in the early 2000 in the dynamics of such systems, the objective being the replacement of the heuristic MRI pulse sequences used in hospital settings (in vivo), which means replacing in near future the standard industrial software by a new generation of software, producing a double gain: a better image in a shorter time.

Since the Bloch equations describe the evolution of the dynamics of the process with great accuracy and the computed control strategies can be implemented easily, this application provides an ideal platform to test the geometric optimal control framework presented in this volume. Clearly, the theory has to be developed to handle the mathematical problems and the analysis has to be supplemented by the use of a new generation of specific software dedicated to optimal control (HamPath [32], Bocop [19], GloptiPoly [47]). With this in mind, the reader is introduced to two important problems in NMR and MRI. The first one is simply to *saturate in minimum time* the magnetization vector, which corresponds to driving its amplitude to zero. For this problem, we must first introduce the most general maximum principle since the applied RF-magnetic field is of bounded amplitude. The second step is to compute, using geometric techniques, the structure of the optimal law as a closed-loop function. This is the so-called concept of optimal synthesis. The second problem is the *contrast problem in MRI* where we must distinguish within a given picture between two heterogeneously distributed species, e.g., healthy versus cancer cells, that are characterized thanks to the Bloch equation by different responses to the same RF-field due to different relaxation parameters. The actual MRI technology enables the transformation of this observation problem into an optimal control problem of the Mayer form:

$$\mathrm{d}x/\mathrm{d}t(t) = f(x(t), u(t)) \text{ with } |u(t)| \leq M, \quad min_{u(.)} c(x(t_f)),$$

where t_f is a fixed transfer time directly related to the image processing time and the cost function measures the contrast. The dynamics represents the coupling of the two Bloch equations controlled by the same RF-field including the respective parameters associated with the two species to be distinguished, parameters which can be easily computed experimentally.

We use three numerical software based on different approaches:

- *Bocop*. The so-called direct approach transforms the infinite-dimensional optimal control problem into a finite-dimensional optimization problem. This is done by a discretization in time applied to the state and control variables, as well as to the dynamics equation. These methods are usually less precise than indirect methods based on the maximum principle, but more robust with respect to the initialization.

- *HamPath*. The HamPath software is based upon indirect methods: simple and multiple shooting; differential continuation (or homotopy) methods; and computation of the solutions of the variational equations needed to check the second-order conditions of local optimality. Shooting methods consist in finding a zero of a specific function and use Newton-like algorithms. While simple shooting leads to solution of a two-point boundary value problem, multiple shooting takes into account intermediate conditions and the structure of the optimal solution has to be determined. This can be done using the Bocop software, which also allows initialization of the shooting equation. The Jacobian of the homotopic function is computed using variational equations to calculate the Jacobi fields. Moreover, the Jacobi fields are used to check the necessary second-order optimality conditions.

- *LMI*. The direct and indirect methods provide local optimal solutions. By comparing the different paths of zeros, one obtains a strong candidate solution whose global optimality must be analyzed. This can be done by the moment approach. The moment approach is a global optimization technique that transforms a nonlinear, possibly infinite-dimensional optimization problem into convex and finite-dimensional relaxations in the form of linear matrix inequalities (LMIs). The first step consists in linearizing the problem by formulating it as a linear program on a measure space, a problem often referred to as a generalized moment problem. This can be performed by the use of so-called occupation measures, encoding admissible trajectories. The second step is to exploit the problem's structure, here given by its polynomial data, to manipulate the measures by their moment sequences. This leads to a semi-definite program (SDP) with countably many decision variables, one for each moment. The third and last step is to truncate this last problem to a finite set of those moments, leading to a relaxation in the form of LMI. The method is used through the `GloptiPoly` software. This approach is developed in the MRI problem thanks to the algebraic structure of Bloch equations and is crucial in this problem to discriminate the global optimum from the multiple local optimum solutions.

Numerical methods are supplemented by *symbolic computations* to handle or to check more complicated calculations. The combination of geometric, numerical, and symbolic computations represents the main originality of the book and leads to the development of a modern and non-trivial computational framework.

Another originality of the work presented here is its connection to real experiments. For the swimming problem, the copepod represents a variety of zooplankton observed at the University of Hawaii in Prof. Takagi's laboratory and is a model for the design of swimming robots. We represent in Fig. 1 the copepod observed by Takagi and the associated micro-robot model.

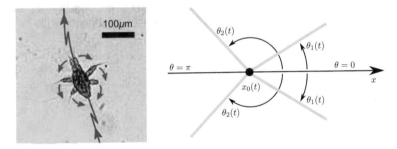

Fig. 1 *Left*: Observation of a zooplankton. *Right*: Sketch of the two-link symmetric swimmer

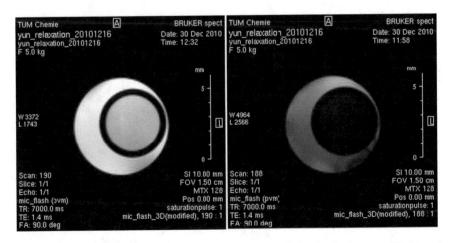

Fig. 2 Experimental results: The inner circle-shape sample mimics the deoxygenated blood, and the outside moon-shape sample corresponds to the oxygenated blood. *Left*: Without control. *Right*: Optimized contrast

Fig. 3 Contrast optimization in an in vivo setting. Species: brain—parietal muscle

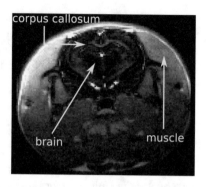

For the MRI problem, the numerical computations were implemented by Prof. Glaser at UTM in in vitro experiments, and more recently, in vivo experiments were performed at Creatis (INSA Lyon) by the group of Prof. Ratiney. Figures 2 and 3 represent the in vivo and in vitro experiments realized in the project. Note that the numerical computations were performed using the Grape algorithm [59].

Dijon, France

Dijon, France

Honolulu, HI, USA

May 2018

Bernard Bonnard

Jérémy Rouot

Monique Chyba

Acknowledgements

- B. Bonnard is partially supported by the ANR Project—DFG Explosys (Grant No. ANR-14-CE35-0013-01;GL203/9-1). M. Chyba is partially supported by the Simons Foundation with Collaboration Grant for Mathematicians: *Geometric Optimal Control and Its Application*. J. Rouot is supported by the European Research Council (ERC) through an ERC Advanced Grant for the TAMING project.
- This book benefited from the support of the FMJH PGMO and from the support of EDF, Thales, and Orange.
- The authors are indebted to Olivier Cots, Alain Jacquemard, Pierre Martinon.
- Many thanks to E. M. Purcell for his scientific work.
- Finally, the first author thanks V. Pécresse for the French 2007 LRU law and the subsequent research and teaching working conditions at UBFC.

Contents

About the Authors

Dr. Bernard Bonnard received his Ph.D in Mathematics from Metz University (France) in 1978 and his Thèse d'Etat degree from Grenoble, France in 1983. He is currently a Professor of Mathematics at the Engineering school Esirem in Dijon, France and member of the INRIA team Mc Tao, Sophia Antipolis. His main research interest is the development of computational techniques in optimal control with applications in space and quantum mechanics.

Dr. Monique Chyba received her Ph.D. in Mathematics from Geneva University (Geneva, Switzerland) in 1997. She is currently a Professor at the department of Mathematics at the University of Hawaii-Manoa, USA. Her main research interest is the development of geometric methods to solve optimal control problems with her central objective being to understand the role of singular extremals in optimal strategies for nonlinear control systems. Her latest work deals with biological applications such as efficient swimming at low Reynolds number and morphogenesis.

Dr. Jérémy Rouot received his Ph.D. in Applied Mathematics from Nice Sophia Antipolis University (France) in 2016. He is currently a lecturer and researcher at the EPF Graduate School of Engineering in Troyes, France. His research focuses on geometric and numerical methods to solve optimal control problems arising from orbital transfer in spatial mechanics, swimming problem at low Reynolds number, nuclear magnetic resonance and biomechanical models.

Chapter 1
Historical Part—Calculus of Variations

The calculus of variations is an old mathematical discipline and historically finds its origins in the introduction of the brachistochrone problem at the end of the 17th century by Johann Bernoulli to challenge his contemporaries to solve it. Here, we briefly introduce the reader to the main results. First, we introduce the fundamental formula of the classical calculus of variations following the presentation by Gelfand Fomin in [41]. The originality of this presentation lies in the fact that it provides a general formula rather than starting with the standard Euler-Lagrange formula derivation and dealing with general variations. The fundamental formula leads to a derivation of the standard first order necessary conditions: Euler-Lagrange equation, tranversality conditions, Erdmann-Weierstrass conditions for a broken extremal and the Hamilton Jacobi equation. Second, we present a derivation of the second order necessary conditions in relation with the concept of conjugate points and the Jacobi equation. The main idea is to introduce the so-called accessory problem replacing the positivity test of the second order derivative by a minimization problem of the associated quadratic form [41]. The modern interpretation in terms of the spectral theory of the associated self-adjoint operator (Morse theory) is also stated.

1.1 Statement of the Problem in the Holonomic Case

We consider the set $\mathscr{C}$ of all curves $x : [t_0, t_1] \to \mathbb{R}^n$ of class C^2, where the initial and final times t_0, t_1 are not fixed, and the problem of minimizing a functional over $\mathscr{C}$:

$$C(x) = \int_{t_0}^{t_1} L(t, x(t), \dot{x}(t)) \, dt$$

© The Author(s), under exclusive license to Springer International Publishing AG, part of Springer Nature 2018
B. Bonnard et al., *Geometric and Numerical Optimal Control*, SpringerBriefs in Mathematics, https://doi.org/10.1007/978-3-319-94791-4_1

where L is C^2. Moreover, we can impose extremity constraints: $x(t_0) \in M_0, x(t_1) \in M_1$ where M_0, M_1 are C^1-submanifolds of $\mathbb{R}^n$. The distance between two curves $x(.), x^*(.)$ is given by

$$\rho(x, x^*) = \max_{t \in J \cap J^*} \|x(t) - x^*(t)\| + \max_{t \in J \cap J^*} \|\dot{x}(t) - \dot{x}^*(t)\| + d(P_0, P_0^*) + d(P_1, P_1^*)$$

where $P_0 = (t_0, x_0)$, $P_1 = (t_1, x_1)$, J, J^* are the domains of x, x^* and $\|\cdot\|$ is any norm on $\mathbb{R}^n$ while d is the usual distance mapping on $\mathbb{R}^{n+1}$. Note that a curve is interpreted in the time-extended space (t, x). If the two curves $x(\cdot), x^*(\cdot)$ are closed, they are by convention C^2-extended on $J \cup J^*$.

Proposition 1 (Fundamental formula of the classical calculus of variations) *We adopt the standard notation of classical calculus of variations, see [41]. Let $\gamma(\cdot)$ be a reference curve with extremities $(t_0, x_0), (t_1, x_1)$ and let $\bar{\gamma}(\cdot)$ be any curve with extremities $(t_0 + \delta t_0, x_0 + \delta x_0), (t_1 + \delta t_1, x_1 + \delta x_1)$. We denote by $h(\cdot)$ the variation: $h(t) = \bar{\gamma}(t) - \gamma(t)$. Then, if we set $\Delta C = C(\bar{\gamma}) - C(\gamma)$, we have*

$$\Delta C = \int_{t_0}^{t_1} \left(\frac{\partial L}{\partial x} - \frac{d}{dt} \frac{\partial L}{\partial \dot{x}} \Big|_{\gamma} \right) \cdot h(t)\, dt + \left[\frac{\partial L}{\partial \dot{x}} \Big|_{\gamma} \cdot \delta x \right]_{t_0}^{t_1}$$

$$+ \left[\left(L - \frac{\partial L}{\partial \dot{x}} \cdot \dot{x} \right)_{\Big|_{\gamma}} \delta t \right]_{t_0}^{t_1} + o(\rho(\bar{\gamma}, \gamma)) \tag{1.1}$$

where $\cdot$ denotes the scalar product in $\mathbb{R}^n$.

Proof We write

$$\Delta C = \int_{t_0 + \delta t_0}^{t_1 + \delta t_1} L(t, \gamma(t) + h(t), \dot{\gamma}(t) + \dot{h}(t))\, dt - \int_{t_0}^{t_1} L(t, \gamma(t), \dot{\gamma}(t))\, dt$$

$$= \int_{t_0}^{t_1} L(t, \gamma(t) + h(t), \dot{\gamma}(t) + \dot{h}(t))\, dt - \int_{t_0}^{t_1} L(t, \gamma(t), \dot{\gamma}(t))\, dt$$

$$+ \int_{t_1}^{t_1 + \delta t_1} L(t, \gamma(t) + h(t), \dot{\gamma}(t) + \dot{h}(t))\, dt - \int_{t_0}^{t_0 + \delta t_0} L(t, \gamma(t) + h(t), \dot{\gamma}(t) + \dot{h}(t))\, dt.$$

We develop this expression using Taylor expansions keeping only the linear terms in $h, \dot{h}, \delta x, \delta t$. We get

$$\Delta C = \int_{t_0}^{t_1} \left(\frac{\partial L}{\partial x} \Big|_{\gamma} \cdot h(t) + \frac{\partial L}{\partial \dot{x}} \Big|_{\gamma} \cdot \dot{h}(t) \right)_{\Big|_{\gamma}} dt + \left[L(t, \gamma, \dot{\gamma}) \delta t \right]_{t_0}^{t_1} + o(h, \dot{h}, \delta t).$$

The derivative of the variation $\dot{h}$ is depending on h, and integrating by parts we obtain

$$\Delta C \sim \int_{t_0}^{t_1} \left(\frac{\partial L}{\partial x} - \frac{d}{dt}\frac{\partial L}{\partial \dot{x}} \right)_{|\gamma} \cdot h(t)\, dt + \left[\frac{\partial L}{\partial \dot{x}}_{|\gamma} \cdot h(t) \right]_{t_0}^{t_1} + \left[L_{|\gamma}\, \delta t \right]_{t_0}^{t_1}.$$

We observe that h, δx, δt are not independent at the extremities and we have for $t = t_0$ or $t = t_1$ the relation

$$h(t + \delta t) = h(t) + o(\dot{h}, \delta t).$$

So

$$h(t) \sim \delta x - \dot{x}\delta t.$$

Hence, we obtain the following approximation:

$$\Delta C \sim \int_{t_0}^{t_1} \left(\frac{\partial L}{\partial x} - \frac{d}{dt}\frac{\partial L}{\partial \dot{x}} \right)_{|\gamma} \cdot h(t)\, dt + \left[\frac{\partial L}{\partial \dot{x}}_{|\gamma} \cdot \delta x \right]_{t_0}^{t_1} + \left[\left(L - \frac{\partial L}{\partial \dot{x}}\dot{x} \right)_{|\gamma} \delta t \right]_{t_0}^{t_1}$$

where all the quantities are evaluated along the reference trajectory $\gamma(\cdot)$. In this formula h, δx, δt can be taken independent because in the integral the values $h(t_0)$, $h(t_1)$ do not play any special role.   □

From (1.1), we deduce that the standard first-order necessary conditions of the calculus of variations.

Corollary 1 *Let us consider the minimization problem where the extremities (t_0, x_0), (t_1, x_1) are fixed. Then, a minimizer $\gamma(\cdot)$ must satisfy the Euler-Lagrange equation*

$$\left(\frac{\partial L}{\partial x} - \frac{d}{dt}\frac{\partial L}{\partial \dot{x}} \right)_{|\gamma} = 0. \tag{1.2}$$

Proof Since the extremities are fixed we set in (1.1) $\delta x = 0$ and $\delta t = 0$ at $t = t_0$ and $t = t_1$. Hence

$$\Delta C = \int_{t_0}^{t_1} \left(\frac{\partial L}{\partial x} - \frac{d}{dt}\frac{\partial L}{\partial \dot{x}} \right)_{|\gamma} \cdot h(t)\, dt + o(h, \dot{h})$$

for each variation $h(\cdot)$ defined on $[t_0, t_1]$ such that $h(t_0) = h(t_1) = 0$. If $\gamma(\cdot)$ is a minimizer, we must have $\Delta C \geq 0$ for each $h(\cdot)$ and clearly by linearity, we have

$$\int_{t_0}^{t_1} \left(\frac{\partial L}{\partial x} - \frac{d}{dt}\frac{\partial L}{\partial \dot{x}} \right)_{|\gamma} \cdot h(t)\, dt = 0$$

for each $h(\cdot)$. Since the mapping $t \mapsto (\frac{\partial L}{\partial x} - \frac{d}{dt}\frac{\partial L}{\partial \dot{x}})_{|\gamma}$ is continuous, it must be identically zero along $\gamma(\cdot)$ and the Euler-Lagrange Equation (1.2) is satisfied.   □

1.2   Hamiltonian Equations

The Hamiltonian formalism, which is the natural formalism to use for the maximum principle, appears in the classical calculus of variations via the Legendre transformation.

Definition 1 The Legendre transformation is defined by

$$p = \frac{\partial L}{\partial \dot{x}}(t, x, \dot{x}) \tag{1.3}$$

and if the mapping $\varphi : (x, \dot{x}) \mapsto (x, p)$ is a diffeomorphism we can introduce the Hamiltonian:

$$H : (t, x, p) \mapsto p \cdot \dot{x} - L(t, x, \dot{x}). \tag{1.4}$$

Remark 1 In mechanics, the Lagrangian L is of the form $V(x) - T(x, \dot{x})$ where V is the potential and T is the kinetic energy and T is strictly convex with respect to $\dot{x}$.

Proposition 2 *The formula (1.1) takes the form*

$$\Delta C \sim \int_{t_0}^{t_1} \left(\frac{\partial L}{\partial x} - \frac{d}{dt} \frac{\partial L}{\partial \dot{x}} \right)_{|_\gamma} \cdot h(t)\, dt + \left[p \delta x - H \delta t \right]_{t_0}^{t_1} \tag{1.5}$$

and if $\gamma(\cdot)$ is a minimizer it satisfies the Euler-Lagrange equation in the Hamiltonian form

$$\dot{x}(t) = \frac{\partial H}{\partial p}(t, x(t), p(t)), \qquad \dot{p}(t) = -\frac{\partial H}{\partial x}(t, x(t), p(t)). \tag{1.6}$$

Proof Computing, one has

$$dH = \frac{\partial H}{\partial t}\, dt + \frac{\partial H}{\partial p}\, dp + \frac{\partial H}{\partial x}\, dx$$

$$= (p - \frac{\partial L}{\partial \dot{x}})\, d\dot{x} + \dot{x}\, dp - \frac{\partial L}{\partial x}\, dx - \frac{\partial L}{\partial t}\, dt.$$

1.3   Hamilton-Jacobi-Bellman Equation

Definition 2 A solution of the Euler-Lagrange equation is called an extremal. Let $P_0 = (t_0, x_0)$ and $P_1 = (t_1, x_1)$. The Hamilton-Jacobi-Bellman (HJB) function is the multivalued function defined by

$$S(P_0, P_1) = \int_{t_0}^{t_1} L(t, \gamma(t), \dot{\gamma}(t))\, dt$$

where $\gamma(\cdot)$ is any extremal with fixed extremities x_0, x_1. If $\gamma(\cdot)$ is a minimizer, it is called the value function.

Proposition 3 *Assume that for each P_0, P_1 there exists a unique extremal joining P_0 to P_1 and suppose that the HJB function is C^1. Let P_0 be fixed and let $\bar{S} : P \mapsto S(P_0, P)$. Then, $\bar{S}$ is a solution of the Hamilton-Jacobi-Bellman equation*

$$\frac{\partial S}{\partial t}(P_0, P) + H(t, x, \frac{\partial S}{\partial x}) = 0. \tag{1.7}$$

Proof Let $P = (t, x)$ and $P + \delta P = (t + \delta t, x + \delta x)$. Denote by $\gamma(\cdot)$ the extremal joining P_0 to P and by $\bar{\gamma}(\cdot)$ the extremal joining P_0 to $P + \delta P$. We have

$$\Delta \bar{S} = \bar{S}(t + dt, x + dx) - \bar{S}(t, x) = C(\bar{\gamma}) - C(\gamma)$$

and from (1.5) it follows that:

$$\Delta \bar{S} = \Delta C \sim \int_{t_0}^{t} \left(\frac{\partial L}{\partial x} - \frac{d}{dt}\frac{\partial L}{\partial \dot{x}} \right)_{|\gamma} \cdot h(t)\, dt + \left[p\delta x - H\delta t \right]_{t_0}^{t},$$

where $h(\cdot) = \bar{\gamma}(\cdot) - \gamma(\cdot)$. Since $\gamma(\cdot)$ is a solution of the Euler-Lagrange equation, the integral is zero and

$$\Delta \bar{S} = \Delta C \sim \left[p\delta x - H\delta t \right]_{t_0}^{t}.$$

In other words, we have

$$d\bar{S} = p\mathrm{d}x - H\, dt.$$

Identifying, we obtain

$$\frac{\partial \bar{S}}{\partial t} = -H, \qquad \frac{\partial \bar{S}}{\partial x} = p. \tag{1.8}$$

Hence we get the HJB equation. Moreover p is the gradient to the level sets $\{x \in \mathbb{R}^n; \ \bar{S}(t, x) = c\}$. □

Other consequences of the general formula are the so-called transversality and Erdmann Weierstrass (1877) conditions. They are presented in the exercises below.

Exercise 1.1 Consider the following problem: among all smooth curves $t \to x(t)$ whose extremity point $P_1 = (t_1, x_1)$ lies on a curve $y(t) = \Psi(t)$, find the curve for which the functional $\int_{t_0}^{t_1} L(t, x, \dot{x})\, dt$ has an extremum. Deduce from the general formula the transversality conditions

$$L + L_{\dot{x}}(\dot{\Psi} - \dot{x}) = 0 \text{ at } t = t_1.$$

Exercise 1.2 Let $t \to x(t)$ be a minimizing solution of $\int_{t_0}^{t_1} L(t, x, \dot{x})\, dt$ with fixed extremities. Assume that $t \to x(t)$ is a broken curve with a corner at $t = c \in]t_0, t_1[$. Prove the Erdmann Weierstrass condition

$$L_{\dot{x}}(c-) = L_{\dot{x}}(c+),$$
$$[L - L_{\dot{x}}\dot{x}](c-) = [L - L_{\dot{x}}\dot{x}](c+).$$

Give an interpretation using Hamiltonian formalism.

1.4   Second Order Conditions

The Euler-Lagrange equation has been derived using the linear terms in the Taylor expansion of ΔC. Using the quadratic terms we can get necessary and sufficient second order conditions. For the sake of simplicity, from now on we assume that the curves $t \mapsto x(t)$ belong to $\mathbb{R}$, and we consider the problem with fixed extremities: $x(t_0) = x_0$, $x(t_1) = x_1$. If the map L is taken C^3, the second derivative is computed as follows:

$$
\begin{aligned}
\Delta C &= \int_{t_0}^{t_1} \left(L(t, \gamma(t) + h(t), \dot{\gamma}(t) + \dot{h}(t)) - L(t, \gamma(t), \dot{\gamma}(t)) \right) dt \\
&= \int_{t_0}^{t_1} \left(\frac{\partial L}{\partial x} - \frac{d}{dt}\frac{\partial L}{\partial \dot{x}} \right)_{|\gamma} \cdot h(t)\, dt + \frac{1}{2}\int_{t_0}^{t_1} \left(\frac{\partial^2 L}{\partial x^2}_{|\gamma} h^2(t) + 2\frac{\partial^2 L}{\partial x \partial \dot{x}}_{|\gamma} h(t)\dot{h}(t) \right. \\
&\quad + \left. \frac{\partial^2 L}{\partial \dot{x}^2}_{|\gamma} \dot{h}^2(t) \right) dt + o(h, \dot{h})^2.
\end{aligned}
$$

If $\gamma(t)$ is an extremal, the first integral is zero and the second integral corresponds to the intrinsic second-order derivative $\delta^2 C$, that is:

$$
\delta^2 C = \frac{1}{2}\int_{t_0}^{t_1} \left(\frac{\partial^2 L}{\partial x^2}_{|\gamma} h^2(t) + 2\frac{\partial^2 L}{\partial x \partial \dot{x}}_{|\gamma} h(t)\dot{h}(t) + \frac{\partial^2 L}{\partial \dot{x}^2}_{|\gamma} \dot{h}^2(t) \right) dt. \tag{1.9}
$$

Using $h(t_0) = h(t_1) = 0$, it can be written after an integration by parts as

$$
\delta^2 C = \int_{t_0}^{t_1} \left(P(t)\dot{h}^2(t) + Q(t)h^2(t) \right) dt \tag{1.10}
$$

where

$$
P = \frac{1}{2}\frac{\partial^2 L}{\partial \dot{x}^2}_{|\gamma}, \qquad Q = \frac{1}{2}\left(\frac{\partial^2 L}{\partial x^2} - \frac{d}{dt}\frac{\partial^2 L}{\partial x \partial \dot{x}} \right)_{|\gamma}.
$$

Using the fact that in the integral (1.10) the term $P\dot{h}^2$ is dominating [41], we get the following proposition.

Proposition 4 *If $\gamma(\cdot)$ is a minimizing curve for the fixed extremities problem then it must satisfy the Legendre condition:*

$$\frac{\partial^2 L}{\partial \dot{x}^2}\bigg|_{\gamma} \geq 0. \tag{1.11}$$

1.5 The Accessory Problem and the Jacobi Equation

The intrinsic second-order derivative is given by

$$\delta^2 C = \int_{t_0}^{t_1} \left(P(t)\dot{h}^2(t) + Q(t)h^2(t) \right) dt, \qquad h(t_0) = h(t_1) = 0,$$

where P, Q are as above. It can be written as

$$\delta^2 C = \int_{t_0}^{t_1} \left((P(t)\dot{h}(t))\dot{h}(t) + (Q(t)h(t))h(t) \right) dt$$

and integrating by parts using $h(t_0) = h(t_1) = 0$, we obtain

$$\delta^2 C = \int_{t_0}^{t_1} \left(Q(t)h(t) - \frac{d}{dt}(P(t)\dot{h}(t)) \right) h(t)\, dt.$$

Let us introduce the linear operator $D : h \mapsto Qh - \frac{d}{dt}(P\dot{h})$. Hence, we can write

$$\delta^2 C = (Dh, h) \tag{1.12}$$

where $(,)$ is the usual scalar product on $L^2([t_0, t_1])$. The linear operator D is called the *Euler-Lagrange operator*.

Definition 3 From (1.12), $\delta^2 C$ is a quadratic operator on the set $\mathscr{C}_0$ of C^2-curves $h : [t_0, t_1] \to \mathbb{R}$ satisfying $h(t_0) = h(t_1) = 0, h \neq 0$. The so-called accessory problem is: $\min_{h \in \mathscr{C}_0} \delta^2 C$.

Definition 4 The Euler-Lagrange equation corresponding to the accessory problem is called the Jacobi equation and is given by

$$Dh = 0 \tag{1.13}$$

where D is the Euler-Lagrange operator: $Dh = Qh - \frac{d}{dt}(P\dot{h})$. It is a second-order linear differential operator.

Definition 5 The strong Legendre condition is $P > 0$, where $P = \frac{1}{2} \frac{\partial^2 L}{\partial \dot{x}^2}|_{\gamma}$. If this condition is satisfied, the operator D is said to be nonsingular.

1.6   Conjugate Point and Local Morse Theory

In this section, we present some results from [43] [72].

Definition 6 Let $\gamma(\cdot)$ be an extremal. A solution $J(\cdot) \in \mathscr{C}_0$ of D $J = 0$ on $[t_0, t_1]$ is called a Jacobi curve. If there exists a Jacobi curve along $\gamma(\cdot)$ on $[t_0, t_1]$ the point $\gamma(t_1)$ is said to be conjugate to $\gamma(t_0)$.

Theorem 1 (Local Morse theory [72]) *Let t_0 be fixed and let us consider the Euler-Lagrange operator (indexed by $t > t_0$) D^t defined on the set $\mathscr{C}_0^t$ of curves on $[t_0, t]$ satisfying $h(t_0) = h(t) = 0$. By definition, a Jacobi curve on $[t_0, t]$ corresponds to an eigenvector J^t associated to an eigenvalue $\lambda^t = 0$ of D^t. If the strong Legendre condition is satisfied along an extremal $\gamma : [t_0, T] \to \mathbb{R}^n$, we have a precise description of the spectrum of D^t as follows. There exists $t_0 < t_1 < \cdots < t_s < T$ such that each $\gamma(t_i)$ is conjugate to $\gamma(t_0)$. If n_i corresponds to the dimension of the set of the Jacobi curves J^{t_i} associated to the conjugate point $\gamma(t_i)$, then for any $\tilde{T}$ such that $t_0 < t_1 < \cdots < t_k < \tilde{T} < t_{k+1} < \cdots < T$ we have the identity*

$$n_{\tilde{T}}^- = \sum_{i=1}^{k} n_i \tag{1.14}$$

where $n_{\tilde{T}}^- = \dim\{$linear space of eigenvectors of $D^{\tilde{T}}$ corresponding to strictly negative eigenvalues$\}$. In particular if $\tilde{T} > t_1$ we have

$$\min_{h \in \mathscr{C}_0} \int_{t_0}^{\tilde{T}} (Q(t)h^2(t) + P(t)\dot{h}^2(t)) \, dt = -\infty. \tag{1.15}$$

1.7   From Calculus of Variations to Optimal Control Theory and Hamiltonian Dynamics

An important and difficult problem is to generalize the first and second order conditions from classical calculus of variations to optimal control theory (OCT).

In OCT, the problem is stated as

$$\begin{cases} \frac{dq}{dt} = F(q, u) \\ \min\limits_{u(\cdot)} \int_0^T L(q, u) \, dt \end{cases}$$

with smooth data but the set of admissible controls is the set of bounded measurable mappings valued in a control domain U, thus the set of admissible trajectories is the set of *absolutely continuous curves* $t \to q(\cdot)$. Minimizers are found among the set of *extremals* (q, p, u) solutions of the Hamilton equations

$$\frac{dq}{dt} = \frac{\partial H}{\partial p}, \quad \frac{dp}{dt} = -\frac{\partial H}{\partial q} \tag{1.16}$$

where H is the so-called unmaximized Hamiltonian

$$H(q, p, u) = p \cdot F(q, u)$$

where the controls have to satisfy the maximization condition

$$H(q, p, u) = \max_{v \in U} H(q, p, v). \tag{1.17}$$

Solving this equation leads in general to several true Hamiltonian $H_i(q, p)$, $i = 1, \ldots, k$ and the optimal solution is found by concatenation of trajectories of the vector fields $\overrightarrow{H}_i$'s.

Remark 1.1 OCT is a non trivial extension to the so-called *Lagrange problem* in calculus of variations since there exists no restriction of the control domain.

The concept of conjugate points can be extended in optimal control and is related to losing optimality for some prescribed topology on the set of controls but practical computation is intricate.

A major problem in the analysis is due to bad controllability properties of the so-called abnormal trajectories. This problem stopped further developments of calculus of variations in the forties [30]. It was revived recently in optimal control theory when dealing with SR-geometry and more geometrically investigated, see for instance [22].

Also in the frame of Hamiltonian formulation of the Maximum Principle defined by (1.16), (1.17) a bridge is open between Hamiltonian dynamics and variational principles. Indeed Hamiltonian and Lagrangian can be related with some regularity assumptions using the Legendre-Fenchel transform

$$H(q, p) = \max_{v} (p \cdot v - L(x, v))$$

and interaction between Hamiltonian dynamics and optimal control is a rich and active domain, see for instance [1, 7].

Chapter 2
Weak Maximum Principle and Application to Swimming at Low Reynolds Number

2.1 Pre-requisite of Differential and Symplectic Geometry

We refer to [9, 42, 46] for more details about the general concepts and notations introduced in this section.

Notations. Let M be a smooth (C^∞ or C^ω) connected and second-countable manifold of dimension n. We denote by TM the fiber bundle and by T^*M the cotangent bundle. Let $V(M)$ be the set of smooth vector fields on M and $Diff(M)$ the set of smooth diffeomorphisms.

Definition 7 Let $X \in V(M)$ and let f be a smooth function on M. The Lie derivative is defined as: $L_X f = df(X)$. If $X, Y \in V(M)$, the Lie bracket is given by

$$ad\, X(Y) = [X, Y] = L_Y \circ L_X - L_X \circ L_Y.$$

If $x = (x_1, \ldots, x_n)$ is a local system of coordinates we have:

$$X(x) = \sum_{i=1}^{n} X_i(x) \frac{\partial}{\partial x_i}$$

$$L_X f(x) = \frac{\partial f}{\partial x} X(x)$$

$$[X, Y](x) = \frac{\partial X}{\partial x}(x) Y(x) - \frac{\partial Y}{\partial x}(x) X(x).$$

The mapping $(X, Y) \mapsto [X, Y]$ is $\mathbb{R}$-linear and skew-symmetric. Moreover, the Jacobi identity holds:

$$[X, [Y, Z]] + [Y, [Z, X]] + [Z, [X, Y]] = 0.$$

© The Author(s), under exclusive license to Springer International Publishing AG, part of Springer Nature 2018
B. Bonnard et al., *Geometric and Numerical Optimal Control*, SpringerBriefs in Mathematics, https://doi.org/10.1007/978-3-319-94791-4_2

Definition 8 Let $X \in V(M)$. We denote by $x(t, x_0)$ the maximal solution of the Cauchy problem $\dot{x}(t) = X(x(t))$, $x(0) = x_0$. This solution is defined on a maximal open interval J containing 0. We denote by $\exp tX$ the local one parameter group associated to X, that is: $\exp tX(x_0) = x(t, x_0)$. The vector field X is said to be complete if the trajectories can be extended over $\mathbb{R}$.

Definition 9 Let $X \in V(M)$ and $\varphi \in Diff(M)$. The image of X by φ is $\varphi * X = d\varphi(X \circ \varphi^{-1})$.

We recall the following results.

Proposition 5 *Let $X, Y \in V(M)$ and $\varphi \in Diff(M)$. We have:*

1. *The one parameter local group of $Z = \varphi * X$ is given by:*

$$\exp tZ = \varphi \circ \exp tX \circ \varphi^{-1}.$$

2. $\varphi * [X, Y] = [\varphi * X, \varphi * Y]$.
3. *The Baker-Campbell-Hausdorff (BCH) formula is:*

$$\exp sX \exp tY = \exp \zeta(X, Y)$$

where $\zeta(X, Y)$ belongs to the Lie algebra generated by $[X, Y]$ with:

$$\zeta(X, Y) = sX + tY + \frac{st}{2}[X, Y] + \frac{st^2}{12}[[X, Y], Y] - \frac{s^2 t}{12}[[X, Y], X]$$

$$- \frac{s^2 t^2}{24}[X, [Y, [X, Y]]] + \cdots .$$

The series is converging for s, t small enough in the analytic case.
4. *We have*

$$\exp tX \exp \varepsilon Y \exp -tX = \exp \eta(X, Y)$$

with $\eta(X, Y) = \varepsilon \sum_{k \geq 0} \frac{t^k}{k!} ad^k X(Y)$ and the series converging for ε, t small enough in the analytic case.
5. *The ad-formula is:*

$$\exp tX * Y = \sum_{k \geq 0} \frac{t^k}{k!} ad^k X(Y)$$

where the series is converging for t small enough.

Definition 10 Let V be a $\mathbb{R}$-linear space of dimension $2n$. This space is said to be symplectic if there exists an application $\omega : V \times V \to \mathbb{R}$ which is bilinear, skew-symmetric and nondegenerate, that is: if $\omega(x, y) = 0$ for all $x \in V$, then $y = 0$. Let W be a linear subspace of V. We denote by $W^\perp$ the set

$$W^\perp = \{x \in V; \ \omega(x, y) = 0 \ \forall y \in W\}.$$

The space W is isotropic if $W \subset W^\perp$. An isotropic space is said to be Lagrangian if dim $W = \dim \frac{V}{2}$. Let (V_1, ω_1), (V_2, ω_2) be two symplectic linear spaces. A linear mapping $f : V_1 \to V_2$ is symplectic if $\omega_1(x, y) = \omega_2(f(x), f(y))$ for each $x, y \in V_1$.

Proposition 6 *Let (V, ω) be a linear symplectic space. Then there exists a basis $\{e_1, \ldots, e_n, f_1, \ldots, f_n\}$ called canonical defined by $\omega(e_i, e_j) = \omega(f_i, f_j) = 0$ for $1 \leq i, j \leq n$ and $\omega(e_i, f_j) = \delta_{ij}$ (Kronecker symbol). If J is the matrix $\begin{pmatrix} 0 & I \\ -I & 0 \end{pmatrix}$ where I is the identity matrix of order n, then we can write $\omega(x, y) = \langle Jx, y \rangle$ where $\langle, \rangle$ is the scalar product (in the basis (e_i, f_j)). In the canonical basis, the set of all linear symplectic transformations is represented as the symplectic group defined by $Sp(n, \mathbb{R}) = \{S \in GL(2n, \mathbb{R}); \ S^\top J S = J\}$.*

Definition 11 Let M be a C^∞-manifold of dimension $2n$. A symplectic structure on M is defined by a 2-form ω such that $d\omega = 0$ and such that ω is regular, that is: $\forall x \in M$, ω_x is nondegenerate.

Proposition 7 *For any C^∞-manifold of dimension n, the cotangent bundle T^*M admits a canonical symplectic structure defined by $\omega = d\alpha$ where α is the Liouville form. If $x = (x_1, \ldots, x_n)$ is a coordinate system on M and (x, p) with $(p_1, \ldots, p_n)$ the associated coordinates on T^*M, the Liouville form is written locally as $\alpha = \sum_{i=1}^n p_i dx_i$ and $\omega = d\alpha = \sum_{i=1}^n dp_i \wedge dx_i$.*

Proposition 8 (Darboux) *Let (M, ω) be a symplectic manifold. Then given any point in M, there exists a local system of coordinates called Darboux coordinates, $(x_1, \ldots, x_n, p_1, \ldots, p_n)$ such that ω is given locally by $\sum_{i=1}^n dp_i \wedge dx_i$. (Hence the symplectic geometry is a geometry with no local invariant).*

Definition 12 Let (M, ω) be a symplectic manifold and let X be a vector field on M. We note $i_X \omega$ the interior product defined by $i_X \omega(Y) = \omega(X, Y)$ for any vector field Y on M. Let $H : M \to \mathbb{R}$ a real-valued function. The vector field denoted by $\overrightarrow{H}$ and defined by $i_{\overrightarrow{H}}(\omega) = -dH$ is the Hamiltonian vector field associated to H. If (x, p) is a Darboux coordinate system, then the Hamiltonian vector field is expressed in these coordinates as:

$$\overrightarrow{H} = \sum_{i=1}^n \frac{\partial H}{\partial p_i} \frac{\partial}{\partial x_i} - \frac{\partial H}{\partial x_i} \frac{\partial}{\partial p_i}.$$

Definition 13 Let $F, G : M \to \mathbb{R}$ be two mappings. We denote by $\{F, G\}$ the Poisson-bracket of F and G defined by $\{F, G\} = dF(\overrightarrow{G})$.

Proposition 9 (Properties of the Poisson-bracket)

1. *The mapping $(F, G) \mapsto \{F, G\}$ is bilinear and skew-symmetric.*

2. *The Leibniz identity holds:*

$$\{FG, H\} = G\{F, H\} + F\{G, H\}.$$

3. *In a Darboux coordinate system, we have*

$$\{F, G\} = \sum_{i=1}^{n} \frac{\partial G}{\partial p_i} \frac{\partial F}{\partial x_i} - \frac{\partial G}{\partial x_i} \frac{\partial F}{\partial p_i}.$$

4. *If the Lie bracket is defined by* $[\vec{F}, \vec{G}] = \vec{G} \circ \vec{F} - \vec{F} \circ \vec{G}$, *then its relation with the Poisson bracket is given by:* $[\vec{F}, \vec{G}] = \{F, G\}$.
5. *The Jacobi identity is satisfied:*

$$\{\{F, G\}, H\} + \{\{G, H\}, F\} + \{\{H, F\}, G\} = 0.$$

Definition 14 Let $\vec{H}$ be a Hamiltonian vector field on (M, ω) and $F : M \to \mathbb{R}$. We say that F is a first integral for $\vec{H}$ if F is constant along any trajectory of $\vec{H}$, that is $dF(\vec{H}) = \{F, H\} = 0$.

Definition 15 Let (x, p) be a Darboux coordinate system and $H : M \to \mathbb{R}$. The coordinate x_1 is said to be cyclic if $\frac{\partial H}{\partial x_1} = 0$. Hence $F : (x, p) \mapsto p_1$ is a first integral.

Definition 16 Let M be a n-dimensional manifold and let (x, p) be Darboux coordinates on T^*M. For any vector field X on M we can define a Hamiltonian vector field $\vec{H}_X$ on T^*M by $H(x, p) = \langle p, X(x) \rangle$; $\vec{H}_X$ is called the Hamiltonian lift of X and $\vec{H}_X = X \frac{\partial}{\partial x} - \frac{\partial X}{\partial x}^{\mathsf{T}} p \frac{\partial}{\partial p}$. Each diffeomorphism φ on M can be lifted into a symplectic diffeomorphism $\vec{\varphi}$ on T^*M defined in a local system of coordinates as follows. If $x = \varphi(y)$, then $\vec{\varphi} : (y, q) \mapsto (x, p) = \left(\varphi(y), \frac{\partial \varphi^{-1}}{\partial y}^{\mathsf{T}} q\right)$.

Theorem 2 (Noether) *Let (x, p) be Darboux coordinates on T^*M, X a vector field on M and $\vec{H}_X$ its Hamiltonian lift. We assume $\vec{H}_X$ to be a complete vector field and we denote by φ_t the associated one parameter group. Let $F : T^*M \to \mathbb{R}$ and let us assume that $F \circ \varphi_t = F$ for all $t \in \mathbb{R}$. Then H_X is a first integral for $\vec{F}$.*

Definition 17 Let M be a manifold of dimension $2n + 1$ and let ω be a 2-form on M. Then for all $x \in M$, ω_x is bilinear, skew-symmetric and its rank is $\leq 2n$. If for each x, the rank is $2n$, we say that ω is regular. In this case $\ker \omega$ is of rank one and is generated by an unique vector field X up to a scalar. If α is a 1-form such that $d\alpha$ is of rank $2n$, the vector field associated with $d\alpha$ is called the characteristic vector field of α and the trajectories of X are called the characteristics.

Proposition 10 *On the space $T^*M \times \mathbb{R}$ with coordinates (x, p, t) the characteristics of the 1-form $\sum_{i=1}^{n}(p_i dx_i - H \, dt)$ project onto solutions of the Hamilton equations:*

$$\dot{x}(t) = \frac{\partial H}{\partial p}(x(t), p(t), t), \quad \dot{p}(t) = -\frac{\partial H}{\partial x}(x(t), p(t), t).$$

Definition 18 Let $\varphi : (x, p, t) \mapsto (X, P, T)$ be a change of coordinates on $T^*M \times \mathbb{R}$. If there exist two functions $K(X, P, T)$ and $S(X, P, T)$ such that

$$p \, dx - H \, dt = P \, dX - K \, dT + dS,$$

then the mapping φ is a canonical transformation and S is called the generating function of φ.

Proposition 11 *For a canonical transformation the equations*

$$\dot{x}(t) = \frac{\partial H}{\partial p}(x(t), p(t), t), \quad \dot{p}(t) = -\frac{\partial H}{\partial x}(x(t), p(t), t)$$

transform onto

$$\frac{dX}{dT}(T) = \frac{\partial K}{\partial P}(X(T), P(T), T), \quad \frac{dP}{dT}(T) = -\frac{\partial K}{\partial X}(X(T), P(T), T).$$

If $T = t$, and (x, X) forms a coordinate system, then we have

$$\frac{dX}{dt}(t) = \frac{\partial K}{\partial P}(X(t), P(t), t), \quad \frac{dP}{dt}(t) = -\frac{\partial K}{\partial X}(X(t), P(t), t)$$

with

$$p(t) = \frac{\partial S}{\partial x}(X(t), P(t), t), \quad P(t) = -\frac{\partial S}{\partial X}(X(t), P(t), t),$$

$$H(X(t), P(t), t) = K(X(t), P(t), t) - \frac{\partial S}{\partial t}(X(t), P(t), t).$$

Remark 2.1 (Integrability) Assume that the generating function S is not depending on t. If there exist coordinates such that $K(X, P) = H(x, p)$ is not depending on P, we have $\dot{X}(t) = 0$, $X(t) = X(0)$; hence $P(t) = P(0) + t \frac{\partial K}{\partial X}\big|_{X=X(0)}$. The equations are integrable. With $H(x, p) = K(X)$ we get

$$H(x, \frac{\partial S}{\partial x}) = K(X).$$

Since $X(t) = (X_1(0), \ldots, X_n(0))$ is fixed, if we can integrate the previous equation we get solutions to the Hamilton equations. A standard method is by separating the variables. This is called the Jacobi method to integrate the Hamilton equations. In particular, this leads to a classification of integrable mechanical systems in small dimension, see [56].

Definition 19 A polysystem D is a family $\{V_i; \ i \in I\}$ of vector fields. We denote by the same letter the associated distribution, that is the mapping $x \mapsto \text{span}\{V(x); V \in D\}$. The distribution D is said to be involutive if $[V_i, V_j] \subset D$, for all $V_i, V_j \in D$.

Definition 20 Let D be a polysystem. We design by $D_{L.A.}$ the Lie algebra generated by D, it is constructed recursively as follows:

$$D_1 = \text{span}\{D\},$$
$$D_2 = \text{span}\{D_1 + [D_1, D_1]\},$$
$$\ldots,$$
$$D_k = \text{span}\{D_{k-1} + [D_1, D_{k-1}]\}$$

and $D_{L.A.} = \cup_{k \geq 1} D_k$. By construction the associated distribution $D_{L.A.}$ is involutive. If $x \in M$, we associate the following sequence of integers: $n_k(x) = \dim D_k(x)$.

Definition 21 Consider a control system $\dot{x} = f(x, u)$ on M with $u \in U$. We can associate to this system the polysystem $D = \{f(\cdot, u); \ u \text{ constant}, \ u \in U\}$. We denote by $S_T(D)$ the set

$$S_T(D) = \{\exp t_1 V_1 \cdots \exp t_k V_k; \ k \in \mathbb{N}, \ t_i \geq 0 \text{ and } \sum_{i=1}^{k} t_i = T, \ V_i \in D\}$$

and by $S(D)$ the local semi-group: $\cup_{T \geq 0} S_T(D)$. We denote by $G(D)$ the local group generated by $S(D)$, that is

$$G(D) = \{\exp t_1 V_1 \cdots \exp t_k V_k; \ k \in \mathbb{N}, \ t_i \in \mathbb{R}, \ V_i \in D\}.$$

Properties.

1. The accessibility set from x_0 in time T is:

$$A(x_0, T) = S_T(D)(x_0).$$

2. The accessibility set from x_0 is the orbit of the local semi-group:

$$A(x_0) = S(D)(x_0).$$

Definition 22 We call the orbit of x_0 the set $O(x_0) = G(D)(x_0)$. The system is said to be weakly controllable if for every $x_0 \in M$, $O(x_0) = M$.

2.2 Controllability Results

2.2.1 Sussmann-Nagano Theorem

When the rank condition is satisfied (rank Δ = constant, $\Delta : x \to D_{L.A.}(x)$) we get from the Frobenius theorem a description of all the integral manifolds near x_0. If we only need to construct the leaf passing through x_0 the rank condition is clearly too strong. Indeed, if $D = \{X\}$ is generated by a single vector field X, there exists an integral curve through x_0. For a family of vector fields this result is still true if the vector fields are analytic.

Theorem 3 (Nagano-Sussman Theorem [85]) *Let D be a family of analytic vector fields near $x_0 \in M$ and let p be the rank of $\Delta : x \mapsto D_{L.A.}(x)$ at x_0. Then through x_0 there exists locally an integral manifold of dimension p.*

Proof Let p be the rank of Δ at x_0. Then there exists p vector fields of $D_{L.A.}$: $X_1, \ldots, X_p$ such that span$\{X_1(x_0), \ldots, X_p(x_0)\} = \Delta(x_0)$. Consider the map

$$\alpha : (t_1, \ldots, t_p) \mapsto \exp t_1 X_1 \ldots \exp t_p X_p(x_0).$$

It is an immersion for $(t_1, \ldots, t_p) = (0, \ldots, 0)$. Hence the image denoted by N is locally a submanifold of dimension p. To prove that N is an integral manifold we must check that for each $y \in N$ near x_0, we have $T_y N = \Delta(y)$. This result is a direct consequence of the equalities

$$D_{L.A.}(\exp t X_i(x)) = \mathrm{d} \exp t X_i(D_{L.A.}(x)), \ i = 1, \ldots, p$$

for x near x_0, and t small enough. To show that the previous equalities hold, let $V(x) \in D_{L.A.}(x)$ such that $V(x) = Y(x)$ with $Y \in D_{L.A.}$. By analycity and the ad-formula for t small enough we have

$$(\mathrm{d} \exp t X_i)(Y(x)) = \sum_{k \geq 0} \frac{t^k}{k!} ad^{\,k} X_i(Y)(\exp t X_i(x)).$$

Hence for t small enough, we have

$$(\mathrm{d} \exp t X_i)(D_{L.A.}(x)) \subset D_{L.A.}(\exp t X_i(x)).$$

Changing t to $-t$ we show the second inclusion.

C^∞-Counter Example

To prove the previous theorem we use the following geometric property. Let X, Y be two analytic vector fields and assume $X(x_0) \neq 0$. From the ad-formula, if all the vector fields $ad^{\,k} X(Y), k \geq 0$ are collinear to X at x_0, then for t small enough the vector field Y is tangent to the integral curve $\exp t X(x_0)$.

Hence is is easy to construct a C^∞-counter example using flat C^∞-mappings. Indeed, let us take $f : \mathbb{R} \mapsto \mathbb{R}$ a smooth map such that $f(x) = 0$ for $x \leq 0$ and $f(x) \neq 0$ for $x > 0$. Consider the two vector fields on $\mathbb{R}^2 : X = \frac{\partial}{\partial x}$ and $Y = f(x)\frac{\partial}{\partial y}$. At 0, $D_{L.A.}$ is of rank 1. Indeed, we have $[X, Y](x) = -f'(x)\frac{\partial}{\partial y} = 0$ at 0 and hence $[X, Y](0) = 0$. The same is true for all high order Lie brackets. In this example the rank $D_{L.A.}$ is not constant along $\exp tX(0)$, indeed for $x > 0$, the vector field Y is transverse to this vector field.

2.2.2   Chow-Rashevskii Theorem

Theorem 4 ([36, 79]) *Let D be a C^∞-polysystem on M. We assume that for each $x \in M$, $D_{L.A.}(x) = T_x M$. Then we have*

$$G(D)(x) = G(D_{L.A.}(x)) = M,$$

for each $x \in M$.

Proof Since M is connected it is sufficient to prove the result locally. The proof is based on the BCH-formula. We assume $M = \mathbb{R}^3$ and $D = \{X, Y\}$ with rank $\{X, Y, [X, Y]\} = 3$ at x_0; the generalization is straightforward. Let λ be a real number and consider the map

$$\varphi_\lambda : (t_1, t_2, t_3) \mapsto \exp \lambda X \exp t_3 Y \exp -\lambda X \exp t_2 Y \exp t_1 X(x_0).$$

We prove that for small but nonzero λ, φ_λ is an immersion. Indeed, using the BCH formula we have

$$\varphi_\lambda(t_1, t_2, t_3) = \exp(t_1 X + (t_2 + t_3)Y + \frac{\lambda t_3}{2}[X, Y] + \cdots)(x_0),$$

hence

$$\frac{\partial \varphi_\lambda}{\partial t_1}(0, 0, 0) = X(x_0), \qquad \frac{\partial \varphi_\lambda}{\partial t_2}(0, 0, 0) = Y(x_0),$$

$$\frac{\partial \varphi_\lambda}{\partial t_3}(0, 0, 0) = Y(x_0) + \frac{\lambda}{2}[X, Y](x_0) + o(\lambda).$$

Since $X, Y, [X, Y]$ are linearly independent at x_0, the rank of φ_λ at 0 is 3 for $\lambda \neq 0$ small enough.

2.3   Weak Maximum Principle

We consider the autonomous control system

$$\dot{x}(t) = f(x(t), u(t)), \qquad x(t) \in \mathbb{R}^n, u(t) \in \Omega \tag{2.1}$$

where f is a C^1-mapping. The initial and target sets M_0, M_1 are given and we assume they are C^1-submanifolds of $\mathbb{R}^n$. The control domain is a given subset $\Omega \subset \mathbb{R}^m$. The class of admissible controls $\mathscr{U}$ is the set of bounded measurable maps $u : [0, T(u)] \to \Omega$. Let $u(\cdot) \in \mathscr{U}$ and $x_0 \in \mathbb{R}^n$ be fixed. Then, by the Caratheodory theorem [64], there exists a unique trajectory of (2.1) denoted $x(\cdot, x_0, u)$ such that $x(0) = x_0$. This trajectory is defined on a nonempty subinterval J of $[0, T(u)]$ and $t \mapsto x(t, x_0, u)$ is an absolutely continuous function solution of (2.1) almost everywhere.

To each $u(\cdot) \in \mathscr{U}$ defined on $[0, T]$ with corresponding trajectory $x(\cdot, x_0, u)$ issued from $x(0) = x_0 \in M_0$ defined on $[0, T]$, we assign a cost

$$C(u) = \int_0^T f^0(x(t), u(t)) \, dt \tag{2.2}$$

where f^0 is a C^1-mapping. An admissible control $u^*(\cdot)$ with corresponding trajectory $x^*(\cdot, x_0, u)$ and defined on $[0, T^*]$ such that $x^*(0) \in M_0$ and $x^*(T^*) \in M_1$ is optimal if for each admissible control $u(\cdot)$ with corresponding trajectory $x(\cdot, x_0, u)$ on $[0, T]$, $x(0) \in M_0$ and $x(T) \in M_1$, then

$$C(u^*) \le C(u).$$

The Augmented System

The following remark is straightforward but is geometrically very important to understand the maximum principle. Let us consider $\hat{f} = (f, f_0)$ and the corresponding system on $\mathbb{R}^{n+1}$ defined by the equations $\dot{\hat{x}} = \hat{f}(\hat{x}(t), u(t))$, i.e.:

$$\dot{x}(t) = f(x(t), u(t)), \tag{2.3}$$
$$\dot{x}^0(t) = f^0(x(t), u(t)). \tag{2.4}$$

This system is called the augmented system. Since $\hat{f}$ is C^1, according to the Caratheodory theorem, to each admissible control $u(\cdot) \in \mathscr{U}$ there exists an admissible trajectory $\hat{x}(t, \hat{x}_0, u)$ such that $\hat{x}_0 = (x_0, x^0(0))$, $x^0(0) = 0$ where the added coordinate $x^0(\cdot)$ satisfies $x^0(T) = \int_0^T f^0(x(t), u(t)) \, dt$.

Let us denote by $\hat{A}_{M_0}$ the accessibility set $\cup_{u(\cdot) \in \mathscr{U}} \hat{x}(T, \hat{x}_0, u)$ from $\hat{M}_0 = (M_0, 0)$ and let $\hat{M}_1 = M_1 \times \mathbb{R}$. Then, we observe that an optimal control $u^*(\cdot)$ corresponds to a trajectory $\hat{x}^*(\cdot)$ such that $\hat{x}^* \in \hat{M}_0$ and intersecting $\hat{M}_1$ at a point $\hat{x}^*(T^*)$ where x^0 is minimal. In particular $\hat{x}^*(T)$ belongs to the boundary of the Accessibility set $\hat{A}_{M_0}$.

Related Problems

Our framework is a general setting to deal with a large class of problems. Examples are the following:

1. Nonautonomous systems:

$$\dot{x}(t) = f(t, x(t), u(t)).$$

We add the variable t to the state space by setting $\frac{dt}{ds} = 1, t(s_0) = s_0$.

2. Fixed time problem. If the time domain $[0, T(u)]$ is fixed ($T(u) = T$ for all $u(\cdot)$) we add the variable t to the state space by setting $\frac{dt}{ds} = 1, t(s_0) = s_0$ and we impose the following state constraints on $t : t = 0$ at $s = 0$ and $t = T$ at the free terminal time s.

Some specific problems important for applications are the following.

1. If $f^0 \equiv 1$, then $\min \int_0^T f^0(x(t), u(t)) \, dt = \min T$ and we minimize the time of global transfer.

2. If the system is of the form: $\dot{x}(t) = A(t)x(t) + B(t)u(t)$, where $A(t), B(t)$ are matrices and $C(u) = \int_0^T L(t, x(t), u(t)) \, dt$ where $L(\cdot, x, u)$ is a quadratic form for each t, T being fixed, the problem is called a linear quadratic problem (LQ-problem).

Singular Trajectories and the Weak Maximum Principle

Definition 23 Consider a system of $\mathbb{R}^n : \dot{x}(t) = f(x(t), u(t))$ where f is a C^∞-map from $\mathbb{R}^n \times \mathbb{R}^m$ into $\mathbb{R}^n$. Fix $x_0 \in \mathbb{R}^n$ and $T > 0$. The end-point map (for fixed x_0, T) is the map $E^{x_0, T} : u(\cdot) \in \mathcal{U} \mapsto x(T, x_0, u)$. If $u(\cdot)$ is a control defined on $[0, T]$ such that the corresponding trajectory $x(\cdot, x_0, u)$ is defined on $[0, T]$, then $E^{x_0, T}$ is defined on a neighborhood V of $u(\cdot)$ for the $L^\infty([0, T])$ norm.

First and Second Variations of $E^{x_0, T}$

It is a standard result, see for instance [84], that the end-point map is a C^∞-map defined on a domain of the Banach space $L^\infty([0, T])$. The formal computation of the successive derivatives uses the concept of Gâteaux derivative. Let us explain in details the process to compute the first and second variations.

Let $v(\cdot) \in L^\infty([0, T])$ be a variation of the reference control $u(\cdot)$ and let us denote by $x(\cdot) + \xi(\cdot)$ the trajectory issued from x_0 and corresponding to the control $u(\cdot) + v(\cdot)$. Since f is C^∞, it admits a Taylor expansion for each fixed t:

$$f(x + \xi, u + v) = f(x, u) + \frac{\partial f}{\partial x}(x, u)\xi + \frac{\partial f}{\partial u}(x, u)v + \frac{\partial^2 f}{\partial x \partial u}(x, u)(\xi, v)$$

$$+ \frac{1}{2}\frac{\partial^2 f}{\partial x^2}(x, u)(\xi, \xi) + \frac{1}{2}\frac{\partial^2 f}{\partial u}(x, u)(v, v) + \cdots$$

Using the differential equation we get

$$\dot{x}(t) + \dot{\xi}(t) = f(x(t) + \xi(t), u(t) + v(t)).$$

Hence we can write ξ as: $\delta_1 x + \delta_2 x + \cdots$ where $\delta_1 x$ is linear in v, $\delta_2 x$ is quadratic, etc. and are solutions of the following differential equations:

$$\dot{\delta_1 x} = \frac{\partial f}{\partial x}(x, u)\delta_1 x + \frac{\partial f}{\partial u}(x, u)v \tag{2.5}$$

$$\dot{\delta_2 x} = \frac{\partial f}{\partial x}(x, u)\delta_2 x + \frac{\partial^2 f}{\partial x \partial u}(x, u)(\delta_1 x, v)$$
$$+ \frac{1}{2}\frac{\partial^2 f}{\partial x^2}(x, u)(\delta_1 x, \delta_2 x) + \frac{1}{2}\frac{\partial^2 f}{\partial u^2}(x, u)(v, v). \tag{2.6}$$

Using $\xi(0) = 0$, these differential equations have to be integrated with the initial conditions

$$\delta_1 x(0) = \delta_2 x(0) = 0. \tag{2.7}$$

Let us introduce the following notations:

$$A(t) = \frac{\partial f}{\partial x}(x(t), u(t)), \qquad B(t) = \frac{\partial f}{\partial u}(x(t), u(t)).$$

Definition 24 The system

$$\dot{\delta x}(t) = A(t)\delta x(t) + B(t)\delta u(t)$$

is called the linearized system along $(x(\cdot), u(\cdot))$.

Let $M(t)$ be the fundamental matrix on $[0, T]$ solution almost everywhere of

$$\dot{M}(t) = A(t)M(t), \qquad M(0) = \mathrm{Id}.$$

Integrating (2.5) with $\delta_1 x(0) = 0$ we get the following expression for $\delta_1 x$:

$$\delta_1 x(T) = M(T) \int_0^T M^{-1}(t) B(t) v(t) \, dt. \tag{2.8}$$

This implies the following lemma.

Lemma 1 The Fréchet derivative of $E^{x_0,T}$ at $u(\cdot)$ is given by

$$E'^{x_0,T}(v) = \delta_1 x(T) = M(T) \int_0^T M^{-1}(t) B(t) v(t) \, dt.$$

Definition 25 The admissible control $u(\cdot)$ and its corresponding trajectory $x(\cdot, x_0, u)$ both defined on $[0, T]$ are said to be regular if the Fréchet derivative $E'^{x_0,T}$ is surjective. Otherwise they are called singular.

Proposition 12 *Let $A(x_0, T) = \cup_{u(\cdot) \in \mathcal{U}} x(T, x_0, u)$ be the accessibility set at time T from x_0. If $u(\cdot)$ is a regular control on $[0, T]$, then there exists a neighborhood U of the end-point $x(T, x_0, u)$ contained in $A(x_0, T)$.*

Proof Since $E'^{x_0,T}$ is surjective at $u(\cdot)$, we have using the open mapping theorem that $E^{x_0,T}$ is an open map.

Theorem 5 *Assume that the admissible control $u(\cdot)$ and its corresponding trajectory $x(\cdot)$ are singular on $[0, T]$. Then there exists a vector $p(\cdot) \in \mathbb{R}^n \setminus \{0\}$ absolutely continuous on $[0, T]$ such that (x, p, u) are solutions almost everywhere on $[0, T]$ of the following equations:*

$$\frac{dx}{dt}(t) = \frac{\partial H}{\partial p}(x(t), p(t), u(t)), \qquad \frac{dp}{dt}(t) = -\frac{\partial H}{\partial x}(x(t), p(t), u(t)) \qquad (2.9)$$

$$\frac{\partial H}{\partial u}(x(t), p(t), u(t)) = 0 \qquad (2.10)$$

where $H(x, p, u) = \langle p, f(x, u) \rangle$ is the pseudo-Hamiltonian, $\langle , \rangle$ being the standard inner product.

Proof We observe that the Fréchet derivative is a solution of the linear system

$$\dot{\delta x}(t) = A(t)\delta_1 x(t) + B(t)v(t).$$

Hence, if the pair $(x(\cdot), u(\cdot))$ is singular this system is not controllable on $[0, T]$. We use an earlier proof on controllability to get a geometric characterization of this property. The proof which is the heuristic basis of the maximum principle is given in detail. By definition, since $u(\cdot)$ is a singular control on $[0, T]$ the dimension of the linear space

$$\left\{ \int_0^T M(T)M^{-1}(t)\, B(t)\, v(t)\, dt; \ v(\cdot) \in L^\infty([0, T]) \right\}$$

is less than n. Therefore there exists a row vector $\underline{p} \in \mathbb{R}^n \setminus \{0\}$ such that

$$\underline{p}M(T)M^{-1}(t)B(t) = 0$$

for almost everywhere $t \in [0, T]$. We set

$$p(t) = \underline{p}M(T)M^{-1}(t).$$

By construction $p(\cdot)$ is a solution of the adjoint system

$$\dot{p}(t) = -p(t)\frac{\partial f}{\partial x}(x(t), u(t)).$$

Moreover, it satisfies almost everywhere the following equality:

$$p(t)\frac{\partial f}{\partial u}(x(t), u(t)) = 0.$$

Hence we get the equations (2.9) and (2.10) if $H(x, p, u)$ denotes the scalar product $\langle p, f(x, u)\rangle$.

Geometric Interpretation of the Adjoint Vector

In the proof of Theorem 5 we introduced a vector $p(\cdot)$. This vector is called an adjoint vector. We observe that if $u(\cdot)$ is singular on $[0, T]$, then for each $0 < t \le T$, $u_{|[0,T]}$ is singular and $p(t)$ is orthogonal to the image denoted $K(t)$ of $E'^{x_0,T}$ evaluated at $u_{|[0,t]}$. If for each t, $K(t)$ is a linear space of codimension one then $p(t)$ is unique up to a factor.

The Weak Maximum Principle

Theorem 6 *Let $u(\cdot)$ be a control and $x(\cdot, x_0, u)$ the corresponding trajectory, both defined on $[0, T]$. If $x(T, x_0, u)$ belongs to the boundary of the accessibility set $A(x_0, T)$, then the control $u(\cdot)$ and the trajectory $x(\cdot, x_0, u)$ are singular.*

Proof According to Proposition 12, if $u(\cdot)$ is a regular control on $[0, T]$ then $x(T)$ belongs to the interior of the accessibility set.

Corollary 2 *Consider the problem of maximizing the transfer time for system $\dot{x}(t) = f(x(t), u(t))$, $u(\cdot) \in \mathcal{U} = L^\infty$, with fixed extremities x_0, x_1. If $u^*(\cdot)$ and the corresponding trajectory are optimal on $[0, \tau^*]$, then $u^*(\cdot)$ is singular.*

Proof If $u^*(\cdot)$ is maximizing then $x^*(T)$ must belong to the boundary of the accessibility set $A(x_0, T)$ otherwise there exists $\epsilon > 0$ such that $x^*(T - \epsilon) \in A(x_0, T)$ and hence can be reached by a solution $\underline{x}(\cdot)$ in time $T : x^*(T - \epsilon) = \underline{x}(T)$. It follows that the point $x^*(T)$ can be joined in a time $\hat{T} > T$. This contradicts the maximality assumption.

Corollary 3 *Consider the system $\dot{x}(t) = f(x(t), u(t))$ where $u(\cdot) \in \mathcal{U} = L^\infty$ ($[0, T]$) and the minimization problem: $\min_{u(\cdot) \in \mathcal{U}} \int_0^T L(x(t), u(t))\, dt$, where the extremities x_0, x_1 are fixed as well as the transfer time T. If $u^*(\cdot)$ and its corresponding trajectory are optimal on $[0, T]$, then $u^*(\cdot)$ is singular on $[0, T]$ for the augmented system: $\dot{x}(t) = f(x(t), u(t))$, $\dot{x}^0(t) = L(x(t), u(t))$. Therefore there exists $\hat{p}^*(t) = (p(t), p_0) \in \mathbb{R}^{n+1} \setminus \{0\}$ such that $(\hat{x}^*, \hat{p}^*, u^*)$ satisfies*

$$\dot{\hat{x}}(t) = \frac{\partial \hat{H}}{\partial \hat{p}}(\hat{x}(t), \hat{p}(t), u(t)), \qquad \dot{\hat{p}}(t) = -\frac{\partial \hat{H}}{\partial \hat{x}}(\hat{x}(t), \hat{p}(t), u(t))$$

$$\frac{\partial \hat{H}}{\partial u}(\hat{x}(t), \hat{p}(t), u(t)) = 0 \qquad (2.11)$$

where $\hat{x} = (x, x^0)$ *and* $\hat{H}(\hat{x}, \hat{p}, u) = \langle p, f(x, u) \rangle + p_0 L(x, u)$. *Moreover* p_0 *is a non-positive constant.*

Proof We have that $x^*(T)$ belongs to the boundary of the accessibility set $\hat{A}(\hat{x}_0, T)$. Applying (2.9), (2.10) we get the Eq. (2.11) where $\dot{p}_0 = -\frac{\partial \hat{H}}{\partial x^0} = 0$ since $\hat{H}$ is independent of x^0.

Abnormality

In the previous corollary, $\hat{p}^*(\cdot)$ is defined up to a factor. Hence we can normalize p_0 to 0 or -1 and we have two cases:

Case 1: $u(\cdot)$ is regular for the system $\dot{x}(t) = f(x(t), u(t))$. Then $p_0 \neq 0$ and can be normalized to -1. This is called the normal case (in calculus of variations), see [30].

Case 2: $u(\cdot)$ is singular for the system $\dot{x}(t) = f(x(t), u(t))$. Then we can choose $p_0 = 0$ and the Hamiltonian $\hat{H}$ evaluated along $(x(\cdot), p(\cdot), u(\cdot))$ doesn't depend on the cost $L(x, u)$. This case is called the abnormal case.

2.4   Second Order Conditions and Conjugate Points

In this section we make a brief introduction to the concept of conjugate point in optimal control, in relation with second order conditions, generalizing the similar concepts in calculus of variations presented in Sect. 2.5.7.

The underlying geometric framework is elegant and corresponds to the concept of Lagrangian manifold [73] and singularity of projection of Lagrangian manifold [8, 90]. They can be numerically computed using rank tests on Jacobi fields which is one of the key components of the HamPath code [38]. Also this concept is well known to be related to the zero eigenvalue of self-adjoint operators associated to the intrinsic second order derivative [52].

2.4.1   Lagrangian Manifold and Jacobi Equation

Definition 26 Let (M, ω) be a (smooth) symplectic manifold of dimension $2n$. A regular submanifold L of M of dimension n is called Lagrangian if the restriction of ω to $T_x L \times T_x L$ is zero.

Definition 27 Let L be a Lagrangian submanifold of T^*M and let $\Pi : z = (x, p) \mapsto x$ be the canonical projection. A tangent non zero vector v of L is called vertical if $d\Pi(v) = 0$. We call caustic the set of points x of L such that there exists at least one vertical field.

Definition 28 Let $\overrightarrow{H}$ be a (smooth) Hamiltonian vector field on T^*M, $\varphi_t = \exp t \overrightarrow{H}$ the associated one parameter group, L_0 the fiber $T_x M$ and $L_t = \varphi_t(L_0)$. The set of caustics is called the set of conjugate loci of L.

Definition 29 Let $\overrightarrow{H}$ be a (smooth) Hamiltonian vector field on T^*M and let $z(t) = (x(t), p(t))$ be a reference trajectory of $\overrightarrow{H}$ defined on $[0, T]$. The variational equation

$$\dot{\delta z}(t) = \frac{\partial \overrightarrow{H}}{\partial z}(z(t))\delta z(t)$$

is called Jacobi equation. We called Jacobi field $J(t) = (\delta x(t), \delta p(t))$ a non trivial solution of Jacobi equation. It is said to be vertical at time t if $\delta x(t) = 0$. A time t_c is called conjugated if there exists a Jacobi field vertical at times 0 and t_c and the point $x(t_c)$ is called geometrically conjugate to $x(0)$.

2.4.2 Numerical Computation of the Conjugate Loci Along a Reference Trajectory

Verticality Test

Let $z(t) = (x(t), p(t))$ be a reference trajectory of $\overrightarrow{H}$ and $x_0 = x(0)$. The set of Jacobi fields forms an n-dimensional linear subspace. Let $(e_1, \ldots, e_n)$ be a basis of $T^*_{x_0}M$ and let $J_i(t) = (\delta x_i(t), \delta p_i(t))$, $i = 1, \ldots, n$ the set of Jacobi fields (vertical at $t = 0$), such that $\delta x_i(0) = 0$, $\delta p_i(0) = e_i$. Therefore the time t_c is geometrically conjugate if and only if the rank of

$$d\Pi_{z(t_c)}(J_1(t_c), \ldots, J_n(t_c))$$

is strictly less than n.

2.5 Sub-riemannian Geometry

In this section a quick introduction to sub-Riemannian (SR-geometry) is presented which is the proper geometry framework for the swimming problem at low Reynolds number.

2.5.1 Sub-riemannian Manifold

Definition 30 A sub-Riemannian manifold is a triple (M, D, g) where M is a smooth connected manifold, D is a smooth distribution of rank m on M and g is a riemannian metric on M.

An *horizontal curve* is an absolutely continuous curve $t \to \gamma(t)$, $t \in I$ such that $\dot{\gamma}(t) \in D(\gamma(t))$. The length of a curve γ is defined by $l(\gamma) = \int_I g(\dot{\gamma}(t))^{1/2} \, dt$ and its *energy* is given by $E(\gamma) = 1/2 \int_0^T g(\dot{\gamma}(t)) \, dt$ where the final time T can be fixed at 1.

2.5.2　Controllability

Let $D_1 = D, D_k = D_1 + [D_1, D_{k-1}]$. We assume that there exists for each $x \in M$ an integer $r(x)$, called the *degree of non holonomy*, such that $D_{r(x)} = T_x M$. Moreover at a point $x \in M$, the distribution D is characterized by the *growth vector* $(n_1, n_2, \ldots, n_r)$ where $n_k = \dim D_k(x)$.

2.5.3　Distance

According to Chow's theorem, for each pair $(x, y) \in M$, there exists an horizontal curve $\gamma : [0, 1] \to M$ such that $\gamma(0) = x$, $\gamma(1) = y$. We denote by d the *sub-Riemannian distance* (SR-distance):

$$d(x, y) = \inf_{\gamma} \{l(\gamma); \ \gamma \text{ is an horizontal curve joining } x \text{ to } y\}.$$

2.5.4　Geodesics Equations

According to Maupertuis principle the length minimization problem is equivalent to the energy minimization problem. Additionally if we parametrize the curves by arc-length, then the length minimization problem is equivalent to the time minimization problem.

To compute the geodesics equations it is convenient to minimize the energy E. We proceed for the calculations as follows. We choose a local orthonormal frame $\{F_1, \ldots, F_m\}$ of D, and we consider the minimization problem:

$$\frac{dx}{dt}(t) = \sum_{i=1}^m u_i(t) F_i(x(t)), \qquad \min_{u(.)} \frac{1}{2} \int_0^1 \left(\sum_i u_i^2(t) \right) dt.$$

According to the weak maximum principle (corresponding to a control domain $U = \mathbb{R}^m$) we introduce the pseudo-Hamiltonian:

$$H(x, p, u) = \sum_{i=1}^{m} u_i H_i(x, p) + p_0 \sum_{i=1}^{m} u_i^2$$

where $H_i(x, p) = \langle p, F_i(x) \rangle$ is the Hamiltonian lift of F_i. By homogeneity p_0 can be normalized to 0 or $-\frac{1}{2}$.

Normal Case: $p_0 = -1/2$.

According to the maximum principle the condition $\frac{\partial H}{\partial u} = 0$ leads to $u_i = H_i$. Plugging this last expression for u_i into H leads to the true Hamiltonian in the normal case:

$$H_n(z) = \frac{1}{2} \sum_{i=1}^{m} H_i^2(z)$$

where $z = (x, p)$. A normal extremal is a solution of the Hamiltonian system associated to the true Hamiltonian, and its projection on the state space is called a normal geodesic.

Abnormal Case $p_0 = 0$.

In this case, the maximum principle leads to the conditions $H_i = 0$, $i = 1, \ldots, m$, thus defining implicitly the abnormal curves related to the structure of the distribution D. The solutions are called abnormal extremals, and their projections on the state space are the abnormal geodesics.

Next we introduce the basic definitions related to the analysis of the geodesics equations and generalizing the Riemannian concepts.

Definition 31 Parametrizing the normal geodesics solutions of $\vec{H}_n(z)$ and fixing $x \in M$, the exponential map is defined by $\exp_x : (p, t) \to \Pi(\exp t\vec{H}_n(z))$ where $z = (x, p)$ and Π is the projection $(x, p) \to x$.

Definition 32 Let $x \in M$ be fixed. The set of points at a SR-distance less or equal to r from x form the ball of radius r centered at x and the sphere $S(x, r)$ is formed by the set of points at a distance r from x.

2.5.5 Evaluation of the Sub-riemannian Ball

The computation of the Sub-Riemannian ball (SR-ball), even with small radius is a very complicated task. One of the most important result in SR-geometry is an approximation result about balls of small radius, in relation with the structure of the distribution.

Definition 33 Let $x \in M$, and let f be a germ of a smooth function at x. The multiplicity of f at x is the number $\mu(f)$ defined by:

- $\mu(f) = \min\{n; \text{ there exist } X_1, \ldots, X_n \in D(x) \text{ such that: } (L_{X_1} \circ \cdots \circ L_{X_n} f)(x) \neq 0\}$,

- if $f(x) \neq 0$ then $\mu(f) = 0$, and $\mu(0) = +\infty$.

Definition 34 Let f be a germ of a smooth function at x, f is called privileged at x if we have that $\mu(f)$ is equivalent to $\min\{k; \ df_x(D^k(x)) \neq 0\}$. A coordinate system $\{x_1, \ldots, x_n\} : V \to \mathbb{R}$ defined on an open subset V of x is called privileged if all the coordinates functions x_i, $1 \leq i \leq n$ are privileged at x.

2.5.6 Nilpotent Approximation

Let us fix a privileged coordinate system at $x = (x_1, \ldots, x_n)$, where the weight of x_i is given by $\mu(x_i)$. Each smooth vector field V at x has a formal expansion $V \sim \sum_{j \geq -1} V^j$, where each $V^j = \sum_{i=1}^n P_i^j(x_1, \ldots, x_n) \frac{\partial}{\partial x_i}$ is homogeneous of degree j for the weights associated with the coordinate system, and the weight of $\frac{\partial}{\partial x_i}$ is $-\mu(x_i)$. $P_i^j(x_1, \ldots, x_n)$ is a homogenous polynomial of degree j.

Proposition 13 *Let $\{F_1, \ldots, F_m\}$ be the orthonormal subframe of the distribution D and set $\hat{F}_i = F_i^{-1}$, $i = 1, \ldots, m$ in the formal expansion. Then, the family $\hat{F}_i$ is a first order approximation of $\{F_1, \ldots, F_m\}$ at x since they generate a nilpotent Lie algebra with a similar growth vector. Moreover, for small x it gives the following estimate of the SR-norm $|x| = d(0, x) \asymp |x_1|^{1/w_1} + \cdots |x_n|^{1/w_n}$.*

See [13, 49, 55] for the details of the construction of privileged coordinates. In addition, note that [71] contains also the relation of the integrability issues which is important for the practical implementation.

2.5.7 Conjugate and Cut Loci in SR-Geometry

The standard concepts of conjugate and cut point from Riemannian geometry can be generalized in optimal control and thus in SR-geometry. Consider the SR-problem:

$$\dot{x}(t) = \sum_{i=1}^m u_i(t) F_i(x(t)), \quad \min_{u(.)} \int_0^T \left(\sum_{i=1}^m u_i^2(t) \right)^{1/2} dt.$$

Definition 35 Let $x(.)$ be a reference (normal or abnormal) geodesic defined on $[0, T]$. The time t_c is called the cut time if the reference geodesic stops to be optimal at $t = t_c$, i.e. is no longer optimal for $t > t_c$, and $x(t_c)$ is called the cut point. Taking all geodesics starting from $x_0 = x(0)$, their cut points will form the cut locus $C_{\text{cut}}(x_0)$. The time t_{1c} is called the first conjugate time if it is the first time such that the reference geodesic is no longer optimal for $t > t_{1c}$ for the C^1-topology on the set of curves, and the point $x(t_{1c})$ is called the first conjugate point. Calculated over all geodesics, the set of first conjugate points will form the (first) conjugate locus $C(x_0)$.

An important step is to relate the computation of the geometric conjugate locus (using a test based on Jacobi fields) to the computation of the conjugate locus associated to optimality. It can be done under suitable assumptions in both the normal and the abnormal case [21] but for simplicity we shall restrict ourselves to the normal case.

2.5.8 Conjugate Locus Computation

Using Maupertuis principle, the SR-problem is equivalent to the (parametrized) energy minimization problem:

$$\min_{u(.)} \int_0^T \left(\sum_{i=1}^m u_i^2(t) \right) dt$$

where T is fixed, and we can choose $T = 1$.

Let $H_i(z) = \langle p, F_i(x) \rangle$ and let $H_n(z) = \frac{1}{2} \sum_{i=1}^m H_i^2(z)$ be the Hamiltonian in the normal case. Take a reference normal geodesic $x(.)$ defined on $[0, 1]$ and let $z(.) = (x(.), p(.))$ be a symplectic lift solution of $\vec{H}_n$. Moreover assume that $x(.)$ is strict, which means that it is not a projection of an abnormal curve. Then the following proposition holds.

Proposition 14 *The first conjugate time t_{1c} along $x(.)$ corresponds to the first geometric conjugate point and can be computed numerically using the test of Sect. 2.4.*

2.5.9 Integrable Case

If the geodesic flow is Liouville integrable, then the Jacobi equation is integrable and the conjugate points can be computed using the parametrization of the geodesic curve. This result is a consequence of the following standard lemma from differential geometry.

Lemma 2 *Let $J(t) = (\delta x(t), \delta p(t))$ be a Jacobi curve along $z(t) = (x(t), p(t))$, $t \in [0, 1]$ and vertical at time $t = 0$, i.e. $\delta x(0) = 0$. Let $\alpha(\varepsilon)$ be any curve in $T^*_{x_0} M$ defined by $p(0) + \varepsilon \delta p(0) + o(\varepsilon)$. Then:*

$$J(t) = \frac{d}{d\varepsilon}_{|\varepsilon=0} \exp t H_n(x(0), \alpha(\varepsilon)).$$

2.5.10 Nilpotent Models in Relation with the Swimming Problem

The models in dimension 3 are related to the classification of stable 2-dimensional distributions, see [91], and will be used for the copepod swimmer. See also [31] for the analysis of the Heisenberg case.

Contact case. A point $x_0 \in \mathbb{R}^3$ is a *contact point* of the distribution $D = \mathrm{span}\{F_1, F_2\}$ if $[F_1, F_2](x_0) \notin D(x_0)$ and the growth vector is $(2, 3)$. A normal form at $x_0 \sim 0$ is given by:

$$x = (x_1, x_2, x_3), \quad D = \ker \alpha, \quad \alpha = x_2 dx_1 + dx_3.$$

Observe that

- $d\alpha = dx_2 \wedge dx_1$: Darboux form,
- $\frac{\partial}{\partial x_3}$ is equal to the Lie bracket $[F_1, F_2]$ and is the characteristic direction of $d\alpha$.

This form is equivalent to the so-called *Dido representation*:

$$D = \ker \alpha', \quad \alpha' = dx_3 + (x_1 dx_2 - x_2 dx_1)$$

with

$$D = \mathrm{span}\{F_1, F_2\}, \quad F_1 = \frac{\partial}{\partial x_1} + x_2 \frac{\partial}{\partial x_3}, \quad F_2 = \frac{\partial}{\partial x_2} - x_1 \frac{\partial}{\partial x_3}.$$

If we set $F_3 = \frac{\partial}{\partial x_3}$, we have that $[F_1, F_2] = 2 F_3$ and the corresponding so-called *Heisenberg SR-case* is given by:

$$\dot{x}(t) = \sum_{i=1}^{2} u_i(t) F_i(x(t)), \quad \min_{u(.)} \int_0^T (u_1^2(t) + u_2^2(t)) \, dt.$$

It corresponds to minimizing the Euclidean length of the projection of the curve $t \to x(t)$ on the (x_1, x_2)-plane. Starting from the origin $(0, 0, 0)$, we observe that

$$x_3(T) = \int_0^T (\dot{x}_1(t) x_2(t) - \dot{x}_2(t) x_1(t)) \, dt$$

is proportional to the area swept by the curve $t \to (x_1(t), x_2(t))$. The Heisenberg SR-case is therefore dual to the Dido problem: among the closed curves in the plane with fixed length, find those for which the enclosed area is maximal. The solutions are well known and they are arcs of circles. They can be easily obtained using simple computations as follows. The geodesic equations written in the (x, H) coordinates where $H = (H_1, H_2, H_3)$, $H_i = \langle p, F_i \rangle$, $i = 1, 2, 3$ are given by:

$$\dot{x}_1 = H_1, \quad \dot{x}_2 = H_2, \quad \dot{x}_3 = H_1 x_2 - H_2 x_1,$$
$$\dot{H}_1 = 2H_2 H_3, \quad \dot{H}_2 = -2H_1 H_3, \quad \dot{H}_3 = 0.$$

Since H_3 is constant we can introduce $H_3 = \lambda/2$ with $\lambda \in \mathbb{R}$, and we obtain the equation of a linear pendulum $dotH_1 + \lambda^2 H_1 = 0$. The integration can be done directly since we can observe that:

$$dotx_3 - \frac{\lambda}{2} \frac{d}{dt}(x_1^2 + x_2^2) = 0.$$

Since $\lambda \neq 0$, which can be assumed positive, we obtain the well known parametrization for the geodesics:

$$x_1(t) = \frac{A}{\lambda}(\sin(\lambda t + \varphi) - \sin(\varphi))$$
$$x_2(t) = \frac{A}{\lambda}(\cos(\lambda t + \varphi) - \cos(\varphi))$$
$$x_3(t) = \frac{A^2}{\lambda}t - \frac{A^2}{\lambda^2}\sin(\lambda t)$$

with $A = \sqrt{H_1^2 + H_2^2}$ and φ is the angle of the vector $(\dot{x}_1, -\dot{x}_2)$.
If $\lambda = 0$, the geodesics are straight lines.

Conjugate points. Computations of first conjugate points are straightforward using the parameterization above for the normal geodesics. Only geodesics whose projections are circles have a first conjugate point given by $t_c = 2\pi/\lambda$ which corresponds to the first intersection of the geodesic with the axis Ox_3. Geometrically, it is due to the symmetry of revolution along this axis which produces a one-parameter family of geodesics starting from 0 and intersecting at such point. This point is also a cut point and a geodesic is optimal up to this point (included).

Note that the SR-Heisenberg case will lead to interesting geometric consequences in the swimming problem: the circles projections correspond to the concept of *stroke*. But while this model can provide some insights on optimal swimming, it is too primitive because:

1. The geodesic flow is integrable due to the symmetries and every (x_1, x_2) motion is periodic;
2. The model is quasi-homogeneous where x_1, x_2 are of weight 1 and x_3 is of weight 2.

Martinet case. A point x_0 is a *Martinet point* if at x_0, $[F_1, F_2] \in \mathrm{span}\{F_1, F_2\}$ and at least one Lie bracket $[[F_1, F_2], F_1]$ or $[[F_1, F_2], F_2]$ does not belong to D. Hence the growth vector is $(2, 2, 3)$. Then, there exist local coordinates near x_0 identified to the origin such that:

$$D = \ker \omega, \quad \omega = dx_3 - \frac{x_2^2}{2} dx_1$$

where

$$F_1 = \frac{\partial}{\partial x_1} + \frac{x_2^2}{2} \frac{\partial}{\partial x_3}, \quad F_2 = \frac{\partial}{\partial x_2}, \quad F_3 = [F_1, F_2] = x_2 \frac{\partial}{\partial x_3}.$$

The surface $\Sigma : \det(F_1, F_2, [F_1, F_2]) = 0$ is identified to $x_2 = 0$ and is called the *Martinet surface*. This surface is foliated by abnormal curves which are integral curves of $\frac{\partial}{\partial x_1}$. In particular abnormal curves passing through the origin and parameterized by arc-length corresponds to the curve $t \rightarrow (t, 0, 0)$.

Those two cases are nilpotent Lie algebras associated to nilpotent approximations of the SR-metric in the copepod swimmer and are respectively the Heisenberg case and the Martinet flat case. Also it can be easily checked that this second case leads to integrable geodesic flow using elliptic functions.

2.6 Swimming Problems at Low Reynolds Number

2.6.1 Purcell's 3-Link Swimmer

The 3-link swimmer is modeled by the position of the center of the second stick $\mathbf{x} = (x, y)$ as well as the angle α between the x-axis and the second stick (the orientation of the swimmer). The shape of the swimmer is modeled by the two relative angles θ_1 and θ_2 (see Fig. 2.1). We also denote respectively by L and L_2 the length of the two external arms and central link. In what follows, x' (resp. x'') corresponds to (x, y) (resp. to $(\alpha, \theta_1, \theta_2)$).

Dynamics via Resistive Force Theory.
We approximate the non local hydrodynamic forces exerted by the fluid on the swimmer with local drag forces depending linearly on the velocity. For each $i \in \{1, 2, 3\}$, we denote by $\mathbf{e}_i^\parallel$ and $\mathbf{e}_i^\perp$ the unit vectors parallel and perpendicular to the i-th link, and we also introduce $\mathbf{v}_i(s)$ the velocity of the point at distance s from the extremity of the i-th link, that is:

Fig. 2.1 Purcell's 3-link swimmer

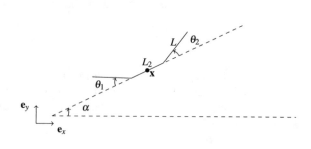

$$\mathbf{v}_1(s) = \dot{\mathbf{x}} - \frac{L_2}{2}\dot{\alpha}\mathbf{e}_2^{\perp} - s(\dot{\alpha} - \dot{\theta}_1)\mathbf{e}_1^{\perp}, \quad s \in [0, L],$$

$$\mathbf{v}_2(s) = \dot{\mathbf{x}} - (s - \frac{L_2}{2})\dot{\alpha}\mathbf{e}_2^{\perp}, \quad s \in [0, L_2],$$

$$\mathbf{v}_3(s) = \dot{\mathbf{x}} + \frac{L_2}{2}\dot{\alpha}\mathbf{e}_2^{\perp} + s(\dot{\alpha} - \dot{\theta}_2)\mathbf{e}_3^{\perp}, \quad s \in [0, L].$$

The force $\mathbf{f}_i$ acting on the i-th segment is taken as:

$$\mathbf{f}_i(s) := -c_t \left(\mathbf{v}_i(s) \cdot \mathbf{e}_i^{\|}\right)\mathbf{e}_i^{\|} - c_n \left(\mathbf{v}_i(s) \cdot \mathbf{e}_i^{\perp}\right)\mathbf{e}_i^{\perp}$$

where c_t and c_n are respectively the drag coefficients in the directions of $\mathbf{e}_i^{\|}$ and $\mathbf{e}_i^{\perp}$. Neglecting inertia forces, Newton laws are written as:

$$\begin{cases} \mathbf{f} = 0, \\ \mathbf{e}_z \cdot \mathbf{T_x} = 0 \end{cases} \tag{2.12}$$

where $\mathbf{f}$ is the total force exerted on the swimmer by the fluid and $\mathbf{e}_z = \mathbf{e}_x \wedge \mathbf{e}_y$,

$$\mathbf{f} = \int_0^L \mathbf{f}_1(s)\,\mathrm{d}s + \int_0^{L_2} \mathbf{f}_2(s)\,\mathrm{d}s + \int_0^L \mathbf{f}_3(s)\,\mathrm{d}s$$

and $\mathbf{T_x}$ is the corresponding total torque computed with respect to the central point $\mathbf{x}$,

$$\mathbf{T_x} = \int_0^L (\mathbf{x}_1(s) - \mathbf{x}_1) \times \mathbf{f}_1(s)\,\mathrm{d}s + \int_0^{L_2} (\mathbf{x}_2(s) - \mathbf{x}_1) \times \mathbf{f}_2(s)\,\mathrm{d}s$$

$$+ \int_0^L (\mathbf{x}_3(s) - \mathbf{x}_1) \times \mathbf{f}_3(s)\,\mathrm{d}s$$

where $x_i = (x_i, y_i)$, for $i = 1, 2, 3$, corresponds to the left-end point of the i-th link, and $x_i(s) = x_i + se_i$.

Since the $\mathbf{f}_i(s)$ are linear in $\dot{\mathbf{x}}$, $\dot{\alpha}$, $\dot{\theta}_1$, $\dot{\theta}_2$, the system (2.12) can be rewritten as

$$A(q) \cdot \begin{pmatrix} \dot{\mathbf{x}} \\ \dot{\alpha} \end{pmatrix} - B(q) \cdot \begin{pmatrix} \dot{\theta}_1 \\ \dot{\theta}_2 \end{pmatrix} = 0$$

where $q(t) = (\theta_1, \theta_2, x, y, \alpha)(t)$. The matrix $A(q)$ is invertible (see [5]). Then, the dynamics of the swimmer is finally expressed as the system

$$\dot{q}(t) = f(q, \dot{\theta}_1, \dot{\theta}_2) = \dot{\theta}_1(t)\, F_1(q(t)) + \dot{\theta}_2(t)\, F_2(q(t))$$

where $\left(F_1(q)\ F_2(q)\right) := \begin{pmatrix} \mathbb{I}_2 \\ A^{-1}(q)B(q) \end{pmatrix}$ with $\mathbb{I}_2$ the 2×2 identity matrix. The equations of the dynamics take the form

$$\begin{pmatrix} \dot{x} \\ \dot{y} \\ \dot{\alpha} \end{pmatrix} = \tfrac{1}{\mathscr{G}}\mathscr{R}_\alpha \begin{pmatrix} g_{11} & g_{12} \\ g_{21} & g_{22} \\ g_{31} & g_{32} \end{pmatrix} \begin{pmatrix} \dot{\theta}_1 \\ \dot{\theta}_2 \end{pmatrix},$$

$$\dot{\theta} = u = S(\theta)\tau \tag{2.13}$$

where τ is the torque, $\mathscr{R}_\alpha$ is the rotation matrix $\mathscr{R}_\alpha = \begin{pmatrix} \cos(\alpha) & -\sin(\alpha) & 0 \\ \sin(\alpha) & \cos(\alpha) & 0 \\ 0 & 0 & 1 \end{pmatrix}$ and g_{ij}, $\mathscr{G}$ and S are functions depending only on (θ_1, θ_2) which have long expressions (cf. [76] for a details).

The cost function u is minimizing the expanded mechanical power

$$\int_0^T \tau \cdot u \, dt \tag{2.14}$$

where $\tau u = u H^{-1} u$ and $H^{-1}(\theta)$ is the symmetric matrix described in [76]. It can be computed as

$$\int_0^T \left(\int_0^L \mathbf{f}_1 \cdot \mathbf{v}_1 + \int_0^{L_2} \mathbf{f}_2 \cdot \mathbf{v}_2 + \int_0^L \mathbf{f}_3 \cdot \mathbf{v}_3 \right).$$

Expressions of the Controlled Vector Fields and the Mechanical Energy.
Normalizing $L = 2$, $L_2 = 1$, $c_t = 1$, $c_n = 2$, we write the swimming control system (2.13) as

$$\dot{q}(t) = \sum_{i=1}^2 u_i(t) F_i(q(t)), \tag{2.15}$$

and we obtain the following expressions of the vector fields F_1, F_2:

$$F_1 = \tfrac{\partial}{\partial \theta_1} + \tfrac{\partial}{\partial x} f_{13} + \tfrac{\partial}{\partial y} f_{14} + \tfrac{\partial}{\partial \alpha} f_{15}, \tag{2.16}$$

$$F_2 = \tfrac{\partial}{\partial \theta_2} + \tfrac{\partial}{\partial x} f_{23} + \tfrac{\partial}{\partial y} f_{24} + \tfrac{\partial}{\partial \alpha} f_{25} \tag{2.17}$$

where
$\delta = 1692 + 336 \cos(\theta_1 - \theta_2) + 84 \cos(2\theta_1) - 24 \cos(\theta_1 + 2\theta_2) - 48 \cos(\theta_1 + \theta_2)$
$+816 \cos(\theta_2) + 72 \cos(-2\theta_2 + \theta_1) + 816 \cos(\theta_1) - 6 \cos(2\theta_1 + 2\theta_2)$
$+18 \cos(-2\theta_2 + 2\theta_1) + 84 \cos(2\theta_2) - 24 \cos(2\theta_1 + \theta_2) + 72 \cos(-\theta_2 + 2\theta_1)$
in

- $f_{13} = 1/\delta \, (4 \sin(\alpha - 2\theta_2) - \sin(\alpha + 2\theta_2 - \theta_1) + 18 \sin(\alpha - \theta_1 - \theta_2)$
  $+3 \sin(\alpha - \theta_1 - 2\theta_2) + 2 \sin(\alpha - 2\theta_1 + 2\theta_2) - 9 \sin(\alpha + \theta_1 - 2\theta_2)$
  $-21 \sin(\alpha + \theta_1 + 2\theta_2) - 126 \sin(\alpha + \theta_1 + \theta_2) - 30 \sin(\alpha - \theta_1 + \theta_2)$
  $-2 \sin(\alpha + 2\theta_1 - 2\theta_2) + 2 \sin(\alpha - 2\theta_1) - 78 \sin(\alpha + \theta_1 - \theta_2)$
  $+16 \sin(\alpha - \theta_2) - 104 \sin(\alpha + \theta_2) - 8 \sin(\alpha + 2\theta_1 - \theta_2) - 24 \sin(\alpha + 2\theta_2)$
  $-18 \sin(\alpha + 2\theta_1) - 36 \sin(\alpha) - 262 \sin(\alpha + \theta_1) + 26 \sin(\alpha - \theta_1)),$

- $f_{14} = 1/\delta \left(18 \cos\left(\alpha + 2\theta_1\right) + 24 \cos\left(\alpha + 2\theta_2\right) + 30 \cos\left(\alpha - \theta_1 + \theta_2\right)\right.$
$-3 \cos\left(\alpha - \theta_1 - 2\theta_2\right) + 126 \cos\left(\alpha + \theta_1 + \theta_2\right) + 78 \cos\left(\alpha + \theta_1 - \theta_2\right)$
$-18 \cos\left(\alpha - \theta_1 - \theta_2\right) + 21 \cos\left(\alpha + \theta_1 + 2\theta_2\right) + 9 \cos\left(\alpha + \theta_1 - 2\theta_2\right)$
$-26 \cos\left(\alpha - \theta_1\right) + 104 \cos\left(\alpha + \theta_2\right) - 16 \cos\left(\alpha - \theta_2\right) + 8 \cos\left(\alpha + 2\theta_1 - \theta_2\right)$
$-4 \cos\left(\alpha - 2\theta_2\right) + 36 \cos\left(\alpha\right) + 262 \cos\left(\alpha + \theta_1\right) + \cos\left(\alpha + 2\theta_2 - \theta_1\right)$
$\left. -2 \cos\left(\alpha - 2\theta_1\right) - 2 \cos\left(\alpha - 2\theta_1 + 2\theta_2\right) + 2 \cos\left(\alpha + 2\theta_1 - 2\theta_2\right)\right)$,
- $f_{15} = 1/\delta \left(-216 - 4 \cos\left(2\theta_1\right) + 6 \cos\left(\theta_1 + 2\theta_2\right) + 12 \cos\left(\theta_1 + \theta_2\right)\right.$
$-204 \cos\left(\theta_1\right) - 18 \cos\left(-2\theta_2 + \theta_1\right) - 84 \cos\left(\theta_1 - \theta_2\right) - 4 \cos$
$\left.(-2\theta_2 + 2\theta_1) + 8 \cos\left(2\theta_2\right)\right)$,
- $f_{23} = 1/\delta \left(21 \sin\left(\alpha + \theta_2 + 2\theta_1\right) - 2 \sin\left(\alpha + 2\theta_1 - 2\theta_2\right) - 2 \sin\left(\alpha - 2\theta_2\right)\right.$
$+9 \sin\left(\alpha + \theta_2 - 2\theta_1\right) + 2 \sin\left(\alpha - 2\theta_1 + 2\theta_2\right) + 30 \sin\left(\alpha + \theta_1 - \theta_2\right)$
$+8 \sin\left(\alpha + 2\theta_2 - \theta_1\right) - 3 \sin\left(\alpha - \theta_2 - 2\theta_1\right) - 18 \sin\left(\alpha - \theta_1 - \theta_2\right)$
$+126 \sin\left(\alpha + \theta_1 + \theta_2\right) + 78 \sin\left(\alpha - \theta_1 + \theta_2\right) + \sin\left(\alpha + 2\theta_1 - \theta_2\right)$
$+262 \sin\left(\alpha + \theta_2\right) + 104 \sin\left(\alpha + \theta_1\right) - 4 \sin\left(\alpha - 2\theta_1\right) - 16 \sin\left(\alpha - \theta_1\right)$
$\left. -26 \sin\left(\alpha - \theta_2\right) + 24 \sin\left(\alpha + 2\theta_1\right) + 18 \sin\left(\alpha + 2\theta_2\right) + 36 \sin\left(\alpha\right)\right)$,
- $f_{24} = 1/\delta \left(4 \cos\left(\alpha - 2\theta_1\right) - 2 \cos\left(\alpha - 2\theta_1 + 2\theta_2\right) - 8 \cos\left(\alpha + 2\theta_2 - \theta_1\right)\right.$
$+2 \cos\left(\alpha - 2\theta_2\right) - 18 \cos\left(\alpha + 2\theta_2\right) + 26 \cos\left(\alpha - \theta_2\right) - 24 \cos\left(\alpha + 2\theta_1\right)$
$-\cos\left(\alpha + 2\theta_1 - \theta_2\right) + 2 \cos\left(\alpha + 2\theta_1 - 2\theta_2\right) - 30 \cos\left(\alpha + \theta_1 - \theta_2\right)$
$-21 \cos\left(\alpha + \theta_2 + 2\theta_1\right) - 126 \cos\left(\alpha + \theta_1 + \theta_2\right) - 78 \cos\left(\alpha - \theta_1 + \theta_2\right)$
$+3 \cos\left(\alpha - \theta_2 - 2\theta_1\right) - 9 \cos\left(\alpha + \theta_2 - 2\theta_1\right) + 18 \cos\left(\alpha - \theta_1 - \theta_2\right)$
$\left. +16 \cos\left(\alpha - \theta_1\right) - 104 \cos\left(\alpha + \theta_1\right) - 262 \cos\left(\alpha + \theta_2\right) - 36 \cos\left(\alpha\right)\right)$,
- $f_{25} = 1/\delta \left(-2168 \cos\left(2\theta_1\right) + 12 \cos\left(\theta_1 + \theta_2\right) + 6 \cos\left(2\theta_1 + \theta_2\right) - 4 \cos\left(2\theta_2\right)\right.$
$\left. -18 \cos\left(2\theta_1 - \theta_2\right) - 204 \cos\left(\theta_2\right) - 4 \cos\left(-2\theta_2 + 2\theta_1\right) - 84 \cos\left(\theta_1 - \theta_2\right)\right)$.

Moreover, writing the integrand of the cost function (2.14) as $au_1^2 + 2bu_1u_2 + cu_2^2$, the coefficients a, b, c are given by

$-a(q) = 1/\kappa \left(3 \cos\left(2\theta_1 + 2\theta_2\right) - 6 \cos\left(-2\theta_2 + 2\theta_1\right) - 12 \cos\left(2\theta_1 - \theta_2\right)\right.$
$\left. +24 \cos\left(2\theta_1 + \theta_2\right) + 72 \cos\left(2\theta_1\right) - 84 \cos\left(2\theta_2\right) - 492 \cos\left(\theta_2\right) - 1233\right)$,

$-b(q) = 1/\kappa \left(\cos\left(2\theta_1 + 2\theta_2\right) - 246 \cos\left(\theta_1\right) - 246 \cos\left(\theta_2\right) + 12 \cos\left(2\theta_1 + \theta_2\right)\right.$
$-6 \cos\left(2\theta_1 - \theta_2\right) + 12 \cos\left(\theta_1 + 2\theta_2\right) + 84 \cos\left(\theta_1 + \theta_2\right) - 276 \cos\left(\theta_1 - \theta_2\right)$
$\left. -6 \cos\left(-2\theta_2 + \theta_1\right) - 4 \cos\left(2\theta_2\right) - 4 \cos\left(2\theta_1\right) - 153\right)$,

$-c(q) = 1/\kappa \left(3 \cos\left(2\theta_1 + 2\theta_2\right) - 492 \cos\left(\theta_1\right) - 6 \cos\left(-2\theta_2 + 2\theta_1\right)\right.$
$+24 \cos\left(\theta_1 + 2\theta_2\right) - 12 \cos\left(-2\theta_2 + \theta_1\right) + 72 \cos\left(2\theta_2\right) - 84 \cos\left(2\theta_1\right) - $
$1233)$.

where $\kappa = 36 \cos\left(\theta_1 - 2\theta_2\right) - 222 \cos\left(2\theta_1\right) - 1116 \cos\left(\theta_2\right) - 222 \cos\left(2\theta_2\right)$
$+18 \cos\left(-2\theta_2 + 2\theta_1\right) - 72 \cos\left(2\theta_1 + \theta_2\right) - 72 \cos\left(\theta_1 + 2\theta_2\right) - 180 \cos$
$(\theta_1 + \theta_2)$
$+36 \cos\left(2\theta_1 - \theta_2\right) - 1116 \cos\left(\theta_1\right) + 36 \cos\left(\theta_1 - \theta_2\right) - 12 \cos\left(2\theta_1 + 2\theta_2\right) - $
3258.

Fig. 2.2 (Symmetric)
copepod swimmer

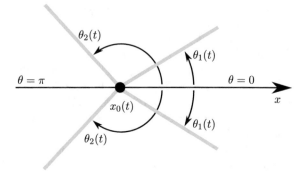

2.6.2 Copepod Swimmer

It is a simplified model proposed by [87] of a symmetric swimming where only
line displacement is authorized, see also [10]. It consists in two pairs of symmetric
links of equal lengths with respective angles θ_1, θ_2 with respect to the displacement
directions Ox while the body is assumed to be an infinitesimal sphere, see Fig. 2.2.

The swimming velocity at x_0 is given by

$$\dot{x}_0 = \frac{\dot{\theta}_1 \sin\theta_1 + \dot{\theta}_2 \sin\theta_2}{2 + \sin^2\theta_1 + \sin^2\theta_2} \tag{2.18}$$

and

$$\dot{\theta}_1 = u_1, \qquad \dot{\theta}_2 = u_2.$$

The mechanical energy is the quadratic form $\dot{q}\, M\, \dot{q}^t$ where $q = (x_0, \theta_1, \theta_2)$ is the
state variable and M is the symmetric matrix

$$M = \begin{pmatrix} 2 - 1/2\,(\cos^2\theta_1 + \cos^2\theta_2) & -1/2\sin\theta_1 & -1/2\sin\theta_2 \\ -1/2\sin\theta_1 & 1/3 & 0 \\ -1/2\sin\theta_2 & 0 & 1/3 \end{pmatrix}.$$

The corresponding Riemannian metric defines the associated SR-metric thanks to
the relation between $\dot{x}_0$ and $\dot{\theta}_1$, $\dot{\theta}_2$.

2.6.3 Some Geometric Remarks

In order to analyze the swimming problem one must introduce the concept of stroke.

Fig. 2.3 Purcell stroke

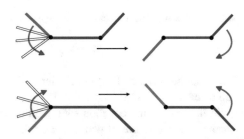

Definition 36 A stroke is a periodic motion of the shape variables associated with a periodic control producing a net displacement of the displacement variable after one period. Observe that due to the SR-structure one can fix the period of the stroke to 2π.

A first geometric analysis is to consider bang-bang controls and the associated strokes. For a single link one gets the famous *scallop theorem*.

Theorem 7 *A scallop cannot swim.*

Proof The relation between the displacement and angular velocity is given by the relation

$$\dot{x}_0 = \frac{\sin(\theta)\dot{\theta}}{2 - \cos^2(\theta)}, \qquad \dot{\theta} = u$$

where θ is the angle of the symmetric link with respect to the axis. Let γ be the angle with respect to the vertical and a stroke is given by

$$u = 1: \qquad \theta : \pi/2 - \gamma \to \pi/2$$
$$u = -1: \qquad \theta : \pi/2 \to \pi/2 - \gamma$$

and the control $u = 1$ produces a displacement: $x_0 \to x_1$ while the control $u = -1$ reverses the motion: $x_1 \to x_0$. The net displacement of the stroke is zero and clearly is related to the reversibility of the SR-model.

A similar computation can be done on the Purcell swimmer using a square stroke like in the original paper ([78]). Considering the controlled system (2.15), the displacement associated with the sequence stroke described in Fig. 2.3 is given by

$$\beta(t) = (\exp tF_2 \, \exp -tF_1 \, \exp -tF_2 \, \exp tF_1)\,(q(0)), \quad q = (\theta_1, \theta_2, x, y, \alpha),$$

and using Baker-Campbell-Hausdorff formula one has

$$\beta(t) = \exp(t^2[F_1, F_2] + o(t^2))(q(0))$$

which gives for small stroke t a displacement of

$$\beta(t) \sim q(0) + t^2[F_1, F_2](q(0)).$$

This shall be compared with [12]. Hence for a small square stroke the displacement can be evaluated using (2.16), (2.17).

In the case of the copepod swimmer, due to the constraints $\theta_i \in [0, \pi]$, $\theta_1 \leq \theta_2$ on the shape variable, a geometric stroke corresponds to a triangle in the shape variable and is defined by $\theta_2 : 0 \to \pi$; $\theta_1 : 0 \to \pi$ and $\theta_1 = \theta_2 : \pi \to 0$. See in the specific analysis of the copepod swimmer the interpretation of this stroke (see Fig. 2.20 (*right*)).

2.6.4   Purcell Swimmer

Due to the mathematical complexity of the expressions of the vector fields F_1 and F_2 (cf. Sect. 2.6.1) employed in this model, the nilpotent approximation will play a crucial role in our analysis. First, as a consequence of the integrability of the associated normal extremals in the class of elliptic functions, the nilpotent approximation will allow us to make a micro-local analysis of the different kinds of strokes and to establish the existence of conjugate points using a suitable time rescaling. Second, the abnormal extremals forming piecewise smooth strokes can be easily computed in this approximation and, then, the optimality of these strokes can be studied using the concept of the (corresponding) conjugate point.

The Flat Nilpotent Model
The Purcell system (2.13) can be written as a control system of the form $\dot{q} = F(q)u = \sum_{i=1}^{2} u_i F_i(q)$, where $q = (\theta_1, \theta_2, x, y, \alpha) \in \mathbb{R}^5$. Even though the vectors fields F_1, F_2 have a complicated expression, they provide a 2-distribution with growth $(2, 3, 5)$ (see [15]). There exists a unique nilpotent model associated with a 2-dimensional distribution in dimension 5 with growth vector $(2, 3, 5)$, see [33, 81].

Definition 2.1 We call the flat Cartan model the 2-dimensional distribution in dimension five defined by the two vector fields:

$$\hat{F}_1(\hat{x}) = \frac{\partial}{\partial \hat{x}_1}, \quad \hat{F}_2(\hat{x}) = \frac{\partial}{\partial \hat{x}_2} + \hat{x}_1 \frac{\partial}{\partial \hat{x}_3} + \hat{x}_3 \frac{\partial}{\partial \hat{x}_4} + \hat{x}_1^2 \frac{\partial}{\partial \hat{x}_5} \tag{2.19}$$

where $\hat{x} = (\hat{x}_1, \hat{x}_2, \hat{x}_3, \hat{x}_4, \hat{x}_5)$ are the privileged coordinates with the following weights: 1 for $\hat{x}_1$ and $\hat{x}_2$, 2 for $\hat{x}_3$, and 3 for $\hat{x}_4$ and $\hat{x}_5$.

Computations of the Nilpotent Approximation
The nilpotent approximation of the Purcell model is computed at the origin. It provides a nilpotent approximation for the SR-problem with the simplified cost

$$\int_0^{2\pi} (u_1^2(t) + u_2^2(t)) \, dt.$$

The two-jets of F_1 and F_2 at $q = (0, 0, 0, 0, 0)$ are given by:

$$F_1(q) = \frac{\partial}{\partial q_1} + \left(-\frac{1}{6} q_5 - \frac{4}{27} q_1 - \frac{2}{27} q_2\right) \frac{\partial}{\partial q_3}$$
$$+ \left(\frac{1}{6} - \frac{1}{12} q_5{}^2 - \frac{2}{27} q_5 q_2 - \frac{4}{27} q_5 q_1 - \frac{1}{27} q_1{}^2 - \frac{1}{27} q_1 q_2 - \frac{1}{36} q_2{}^2\right) \frac{\partial}{\partial q_4}$$
$$+ \left(-\frac{7}{27} + \frac{2}{81} q_1{}^2 - \frac{2}{81} q_1 q_2 - \frac{5}{162} q_2{}^2\right) \frac{\partial}{\partial q_5} + O(|q|^3)$$

$$F_2(q) = \frac{\partial}{\partial q_2} + \left(\frac{1}{6} q_5 + \frac{4}{27} q_2 + \frac{2}{27} q_1\right) \frac{\partial}{\partial q_3}$$
$$+ \left(-\frac{1}{6} + \frac{1}{12} q_5{}^2 + \frac{2}{27} q_5 q_2 + \frac{2}{27} q_5 q_1 + \frac{1}{36} q_1{}^2 + \frac{1}{27} q_1 q_2 + \frac{1}{27} q_2{}^2\right) \frac{\partial}{\partial q_4}$$
$$+ \left(-\frac{7}{27} - \frac{5}{162} q_1{}^2 - \frac{2}{81} q_1 q_2 + \frac{2}{81} q_2{}^2\right) \frac{\partial}{\partial q_5} + O(|q|^3).$$

The local diffeomorphism φ, which transforms F_1, F_2 into the nilpotent approximation $\hat{F}_1, \hat{F}_2$, can be explicitly written using a sequence $\varphi = \varphi_N \, o \cdots o \, \varphi_1 : \mathbb{R}^5 \to \mathbb{R}^5$, where $N = 13$ (see [15]). This leads to a complicated transformation whose role is to relate the privileged coordinates to the physical coordinates $(\theta_1, \theta_2, x, y, \alpha)$ in particular we have a 'stability' property for the shape variables as stated in the next proposition.

Proposition 2.1 *The shape variables* $\theta = (\theta_1, \theta_2)$ *corresponds to the* $(\hat{x}_1, \hat{x}_2)$ *coordinates.*

Integration of Normal Extremal Trajectories
Computing with (2.19), we obtain:

$$\hat{F}_1(\hat{x}) = \frac{\partial}{\partial \hat{x}_1}, \qquad\qquad \hat{F}_2(\hat{x}) = \frac{\partial}{\partial \hat{x}_2} + \hat{x}_1 \frac{\partial}{\partial \hat{x}_3} + \hat{x}_3 \frac{\partial}{\partial \hat{x}_4} + \hat{x}_1^2 \frac{\partial}{\partial \hat{x}_5},$$

$$[\hat{F}_1, \hat{F}_2](\hat{x}) = -\frac{\partial}{\partial \hat{x}_3} - 2\hat{x}_1 \frac{\partial}{\partial \hat{x}_5}, \quad [[\hat{F}_1, \hat{F}_2], \hat{F}_1](\hat{x}) = -2\frac{\partial}{\partial \hat{x}_5},$$

$$[[\hat{F}_1, \hat{F}_2], \hat{F}_2](\hat{x}) = \frac{\partial}{\partial \hat{x}_4}.$$

All brackets of length greater than 3 are zero. Let us introduce $\hat{z} = (\hat{x}, \hat{p})$. Employing the corresponding Hamiltonian lifts, we have:

$$H_1(\hat{z}) = \langle \hat{p}, \hat{F}_1(\hat{x}) \rangle = \hat{p}_1, \quad H_2(\hat{z}) = \langle \hat{p}, \hat{F}_2(\hat{x}) \rangle = \hat{p}_2 + \hat{p}_3 \hat{x}_1 + \hat{p}_4 \hat{x}_3 + \hat{p}_5 \hat{x}_1^2,$$
$$H_3(\hat{z}) = \langle \hat{p}, [\hat{F}_1, \hat{F}_2](\hat{x}) \rangle = -\hat{p}_3 - 2\hat{x}_1 \hat{p}_5, \quad H_4(\hat{z}) = \langle \hat{p}, [[\hat{F}_1, \hat{F}_2], \hat{F}_1](\hat{x}) \rangle = -2\hat{p}_5,$$
$$H_5(\hat{z}) = \langle \hat{p}, [[\hat{F}_1, \hat{F}_2], \hat{F}_2](\hat{x}) \rangle = \hat{p}_4.$$

The SR-Cartan flat case is

$$\dot{\hat{x}}(t) = \sum_{i=1}^{2} u_i(t)\hat{F}_i(\hat{x}(t)), \quad \min_u \int_0^{2\pi} (u_1^2(t) + u_2^2(t))\, dt.$$

and the normal Hamiltonian takes the form

$$H_n = 1/2\,(H_1^2 + H_2^2). \tag{2.20}$$

More precisely, using the Poincaré coordinates, the control system can be written as:

$$\dot{\hat{x}}_1 = H_1, \quad \dot{\hat{x}}_2 = H_2, \quad \dot{\hat{x}}_3 = H_2\hat{x}_1,$$
$$\dot{\hat{x}}_4 = H_2\hat{x}_3, \quad \dot{\hat{x}}_5 = H_2\hat{x}_1^2. \tag{2.21}$$

By differentiating with respect to the time variable, we obtain:

$$\dot{H}_1 = dH_1(\vec{H_n}) = \{H_1, H_2\}H_2 = \langle \hat{p}, [\hat{F}_1, \hat{F}_2](\hat{x})\rangle H_2 = H_3H_2,$$
$$\dot{H}_2 = -H_3H_1, \quad \dot{H}_3 = H_1H_4 + H_2H_5,$$
$$\dot{H}_4 = 0 \quad \text{hence} \quad H_4 = c_4, \quad \dot{H}_5 = 0 \quad \text{hence} \quad H_5 = c_5.$$

We fix the energy level $H_1^2 + H_2^2$ to 1, and we introduce $H_1 = \cos\vartheta$ and $H_2 = \sin\vartheta$ which implies:

$$\dot{H}_1 = -\sin\vartheta\,\dot{\vartheta} = H_2H_3 = \sin\vartheta\,H_3.$$

It follows that $\dot{\vartheta} = -H_3$ and

$$dot\vartheta = -(H_1c_4 + H_2c_5) = -c_4\cos\vartheta - c_5\sin\vartheta = -\omega^2\sin(\vartheta + \phi) \tag{2.22}$$

where ω and ϕ are constant. More precisely, we have:

$$\omega = \left(\hat{p}_{40}^2 + 4\hat{p}_{50}^2\right)^{1/4}, \quad \phi = \arctan\left(-2\hat{p}_{50}/\hat{p}_{40}\right).$$

- First, we consider the *degenerate case* which corresponds to $dot\vartheta = 0$. Therefore, $\vartheta(t) = \hat{p}_{30}\,t + \vartheta_0$ where $\vartheta_0, \hat{p}_{30}$ are constant and for $\hat{p}_{30} \neq 0$, the solutions $\hat{x}_1, \hat{x}_2$ of (2.21) are expressed as:

$$\hat{x}_1(t) = \hat{x}_{10} + 1/\hat{p}_{30}\,\sin(\hat{p}_{30}\,t + \vartheta_0),$$
$$\hat{x}_2(t) = \hat{x}_{20} - 1/\hat{p}_{30}\,\cos(\hat{p}_{30}\,t + \vartheta_0) \tag{2.23}$$

where $\hat{x}_{10}, \hat{x}_{20}$ are constant.

- Second, the case corresponding to $\dot{\vartheta} \neq 0$ leads to a pendulum equation. Indeed, by introducing $\psi = \vartheta + \phi$, (2.22) becomes:

$$1/2 \, \dot{\psi}^2 - \omega^2 \cos(\psi) = B, \qquad (2.24)$$

where B is the constant

$$B = 1/2 \left(\hat{p}_{30} + 2\,\hat{x}_{10}\hat{p}_{50}\right)^2 - \hat{p}_{10}\,\hat{p}_{40} - 2\,\hat{p}_{50}\,\hat{p}_{20} - 2\,\hat{p}_{50}\,\hat{p}_{40}\,\hat{x}_{30}.$$

We have the following two possible cases.

- *Oscillating case.* We introduce $k^2 = 1/2 + B/(2\,\omega^2)$ with $0 < k < 1$ so that (2.24) becomes

$$\dot{\psi}^2 = 4\omega^2 \left(k^2 - \sin^2(\psi/2)\right)$$

and, using standard relations on elliptic functions (cf. [63]), we obtain

$$\sin(\psi/2) = k \, \mathrm{sn}(u, k), \quad \cos(\psi/2) = \mathrm{dn}(u, k)$$

where $u = \omega t + \varphi_0$. cn and dn are elliptic functions of the first kind and the solutions of (2.21), $\hat{x}_1$, $\hat{x}_2$, are expressed as

$$\omega\hat{x}_1(u) = \omega\hat{x}_{10} + -2k \, \sin(\phi) \, \mathrm{cn}(u) + (-u + 2E(u))\cos(\phi)$$
$$\omega\hat{x}_2(u) = \omega\hat{x}_{20} + -2k \, \cos(\phi) \, \mathrm{cn}(u) + (u - 2E(u))\sin(\phi) \quad (2.25)$$

where $\hat{x}_{10}$ and $\hat{x}_{20}$ are constant, and $E(.)$ is the elliptic integral of the second kind.

- *Rotating case.* We introduce $k^2 = 2\,\omega^2/(B + \omega^2)$ with $0 < k < 1$ so that (2.24) becomes

$$\dot{\psi}^2 = 4\omega^2/k^2 \left(1 - k^2 \sin^2(\psi/2)\right).$$

Invoking again elliptic functions properties ([63]) we have

$$\sin(\psi/2) = \mathrm{sn}(u/k, k), \quad \cos(\psi/2) = \mathrm{cn}(u/k, k)$$

where $u = \omega t + \varphi_0$. Still sn and cn are elliptic functions of the first kind. The solutions of (2.21), $\hat{x}_1$, $\hat{x}_2$, satisfy the relations

$$\omega\hat{x}_1(u) = \omega\hat{x}_{10} + \left(1 - \frac{2}{k^2} + 2\frac{E(k)}{k^2\,K(k)}\right)\cos(\phi)\,u + \frac{2}{k}\left(\cos(\phi)\,Z\left(\frac{u}{k}\right) - \sin(\phi)\,\mathrm{dn}\left(\frac{u}{k}\right)\right)$$
$$\omega\hat{x}_2(u) = \omega\hat{x}_{20} + \left(\frac{2}{k^2} - 1 - 2\frac{E(k)}{k^2\,K(k)}\right)\sin(\phi)\,u - \frac{2}{k}\left(\sin(\phi)\,Z\left(\frac{u}{k}\right) + \cos(\phi)\,\mathrm{dn}\left(\frac{u}{k}\right)\right)$$

$$(2.26)$$

where $\hat{x}_{10}$ and $\hat{x}_{20}$ are constant, $K(k)$, $E(k)$ are respectively the complete elliptic integrals of the first and second kind, $Z(.)$ is the Jacobi's Zeta function.

Computations of Strokes with Small Amplitudes using the Nilpotent Approximation

We recall that the physical variables q are related to $\hat{x}$ using the transformation φ. The adjoint variables p are obtained by a Mathieu transformation associated with φ. More precisely, according to Proposition 2.1, recall that the shape variables $\theta = (\theta_1, \theta_2)$ correspond to the $(\hat{x}_1, \hat{x}_2)$ coordinates.

Strokes with small amplitudes such that $q(0) = 0$ are computed from the nilpotent approximation in the following way:

- *Degenerate case*: The corresponding solutions $\hat{x}_i(.)$, $i = 1, 2$ of (2.23) yield the periodic shape variables $\theta_i(t) = \hat{x}_i(t)$, $i = 1, 2$ of period $2\pi/\hat{p}_{30}$. Moreover, the constants $\hat{x}_{10}, \hat{x}_{20}, \vartheta_0$ may be chosen so that $q(0) = (\theta_1(0), \theta_2(0), x(0)) = 0$.
- *Oscillating case*:
  The modulus k can be expressed as

$$k(\hat{p}(0)) = \frac{1}{2}\sqrt{\frac{2\sqrt{\hat{p}_{40}^2 + 4\hat{p}_{50}^2} + \hat{p}_{30}^2 - 2\hat{p}_{10}\hat{p}_{40} - 4\hat{p}_{50}\hat{p}_{20}}{\sqrt{\hat{p}_{40}^2 + 4\hat{p}_{50}^2}}} \qquad (2.27)$$

and, computing $k(\hat{p}(0))$ such that the linear terms of $\theta_1(t) = \hat{x}_1(\omega t + \varphi_0)$, $\theta_2(t) = \hat{x}_2(\omega t + \varphi_0)$ of (2.25) vanish, leads to periodic strokes with eight shapes of period

$$T = 4K(k)/\left(\hat{p}_{40}^2 + 4\hat{p}_{50}^2\right)^{1/4}.$$

The constants $\hat{x}_{10}, \hat{x}_{20}$ are chosen such that $\vartheta(0) = 0$. The initial adjoint vector $\hat{p}(0)$ has to verify the conditions $H_1(\hat{x}(0), \hat{p}(0))^2 + H_2(\hat{x}(0), \hat{p}(0))^2 = 1, k(\hat{p}(0)) \in (0, 1)$ and $\hat{p}_{40}^2 + 4\hat{p}_{50}^2 \neq 0$.
We integrate numerically the stroke in the physical variables starting from $(q(0) = 0, \hat{p}(0))$ and show that the stroke has a conjugate point on $[0, T]$.

- *Rotating case*: The modulus k can be expressed as

$$k(\hat{p}(0)) = 2\sqrt{\frac{\sqrt{\hat{p}_{40}^2 + 4\hat{p}_{50}^2}}{2\sqrt{\hat{p}_{40}^2 + 4\hat{p}_{50}^2} + \hat{p}_{30}^2 - 2\hat{p}_{10}\hat{p}_{40} - 4\hat{p}_{50}\hat{p}_{20}}} \qquad (2.28)$$

We have $\theta_1(t) = \hat{x}_1(\omega t + \varphi_0)$, $\theta_2(t) = \hat{x}_2(\omega t + \varphi_0)$ where $\hat{x}_1, \hat{x}_2$ are explicitly written in (2.26). We choose $p(0)$ so that $H_1(\hat{x}(0), \hat{p}(0))^2 + H_2(\hat{x}(0), \hat{p}(0))^2 = 1$, $k(\hat{p}(0)) \in (0, 1)$ and such that the denominator of $k(\hat{p}(0))$ is nonzero. As $k(\hat{p}(0))$ tends to 0, the linear terms of $\hat{x}_1(u), \hat{x}_2(u)$ of (2.26) tend to 0. This is the case when $\hat{p}_{40} \to 0$ and $\hat{p}_{50} \to 0$, and at the limit, Eq. (2.22) reduces to the equation of the degenerate case: $\dot{\vartheta} = 0$.

Abnormal Case

We can reduce the problem by considering the minimal time problem for the single-input affine system (cf. [21]):

$$\dot{\hat{x}}(t) = \hat{F}_1(\hat{x}(t)) + u(t)\hat{F}_2(\hat{x}(t))$$

where $u(.)$ is now a scalar control. We denote by $\hat{x}(.)$ a reference minimum time trajectory, and since we consider abnormal extremals it follows from the Pontryagin maximum principle that along the extremal lift of $\hat{x}(.)$, the identity $H_2(\hat{x}, \hat{p}) = 0$ must hold and, differentiating with respect to t, it implies that $\{H_1, H_2\}(\hat{x}, \hat{p}) = 0$ must hold too. Differentiating once more time, the extremals associated with the controls:

$$u_a(\hat{x}, \hat{p}) = -\{H_1, \{H_2, H_1\}\}(\hat{x}, \hat{p}) / \{H_2, \{H_1, H_2\}\}(\hat{x}, \hat{p}) = 2\hat{p}_5 / \hat{p}_4$$

satisfy the relation $H_2 = \{H_1, H_2\} = 0$ along $(\hat{x}(.), \hat{p}(.))$ and are solutions of:

$$\dot{\hat{x}}(t) = \frac{\partial H_a}{\partial \hat{p}}(\hat{x}(t), \hat{p}(t)), \quad \dot{\hat{p}}(t) = -\frac{\partial H_a}{\partial \hat{x}}(\hat{x}(t), \hat{p}(t)),$$

where H_a is the true Hamiltonian:

$$H_a(\hat{x}, \hat{p}) = H_1(\hat{x}, \hat{p}) + u_a H_2(\hat{x}, \hat{p}) = \hat{p}_1 + 2\hat{p}_5 \left(\hat{p}_2 + \hat{p}_3 \hat{x}_1 + \hat{p}_4 \hat{x}_3 + \hat{p}_5 \hat{x}_1^2\right)/\hat{p}_4.$$

From the Pontryagin maximum principle, we also have that $H_1(\hat{x}(.), \hat{p}(.)) = 0$. The extremal system subject to the constraints $H_1 = H_2 = \{H_1, H_2\} = 0$ is integrable and the corresponding solutions can be written as:

$$\hat{x}_1(t) = t + \hat{x}_{10}, \quad \hat{x}_2(t) = 2\hat{p}_{50}/\hat{p}_{40}t + \hat{x}_{20},$$

$$\hat{x}_3(t) = \hat{p}_{50}/\hat{p}_{40}t^2 + 2\hat{p}_{50}\hat{x}_{10}/\hat{p}_{40}t + \hat{x}_{30},$$

$$\hat{x}_4(t) = 2/3\,\hat{p}_{50}^2/\hat{p}_{40}^2 t^3 - 2\hat{p}_{50}/\hat{p}_{40}^2 \left(\hat{p}_{50}\hat{x}_{10}^2 + \hat{p}_{30}\hat{x}_{10} + \hat{p}_{20}\right)t$$
$$\quad - \hat{p}_{50}\hat{p}_{30}/\hat{p}_{40}^2 t^2 + \hat{x}_{40},$$

$$\hat{x}_5(t) = 2/3\,\hat{p}_{50}/\hat{p}_{40}t^3 + \left(4\hat{p}_{50}\hat{x}_{10} + \hat{p}_{30}\right)/\hat{p}_{40}t^2$$
$$\quad + 2\left(2\hat{p}_{50}\hat{x}_{10}^2 + \hat{p}_{30}\hat{x}_{10} + \hat{x}_{30}\hat{p}_{40} + \hat{p}_{20}\right)/\hat{p}_{40}t + \hat{x}_{50},$$

$$\hat{p}_1(t) = \left(-2\hat{p}_{50}\hat{p}_{30} - 4\hat{p}_{50}^2\hat{x}_{10}\right)/\hat{p}_{40}t + \hat{p}_{10},$$

$$\hat{p}_2(t) = \hat{p}_{20}, \quad \hat{p}_3(t) = -2\hat{p}_{50}t + \hat{p}_{30}, \quad \hat{p}_4(t) = \hat{p}_{40}, \quad \hat{p}_5(t) = \hat{p}_{50}$$

with $(\hat{x}_{10}, \hat{x}_{20}, \hat{x}_{30}, \hat{x}_{40}, \hat{x}_{50}, \hat{p}_{10}, \hat{p}_{20}, \hat{p}_{30}, \hat{p}_{40}, \hat{p}_{50})$ are constant satisfying

$$\hat{p}_{10} = 0, \quad \hat{p}_{20} = \hat{p}_{50}\hat{x}_{10}^2 - \hat{p}_{40}\hat{x}_{30}, \quad \hat{p}_{30} = -2\hat{p}_{50}\hat{x}_{10}.$$

Remark 2.2 The θ-projection of abnormals are straight lines and form triangular strokes.

2.7 Numerical Results

This section presents the numerical simulations performed on the Purcell swimmer problem. Simulations are performed using both direct and indirect methods, respectively with the solvers `Bocop` and `HamPath`. We use the multipliers from the solutions of the direct method to initialize the adjoint variables in the indirect approach. We display the optimal trajectories obtained for both the nilpotent approximation as well as for the true mechanical system.

BOCOP.

`Bocop` (www.bocop.org, [19]) implements a so-called direct transcription method. More precisely, a time discretization is used to rewrite the optimal control problem as a finite dimensional optimization problem (i.e. nonlinear programming), solved by an interior point method (IPOPT). We recall below the optimal control problem, formulated with the state $q = (\theta_1, \theta_2, x, y, \alpha)$ and control $u = (\dot{\theta}_1, \dot{\theta}_2)$:

$$
\begin{cases}
\min_u \int_0^T E(u(t))\, dt \\
\dot{q}(t) = F_1(q(t))\, u_1(t) + F_2(q(t))\, u_2(t) \\
x(0) = y(0) = 0, \ x(T) = x_f \\
y(T) = y_f, \ \alpha(T) = \alpha(0), \ \theta_i(T) = \theta_i(0), \ i = 1, 2.
\end{cases}
\tag{2.29}
$$

HamPath.

The `HamPath` software (http://www.hampath.org/, [38]) is based on indirect methods to solve optimal control problems using simple shooting methods and testing the local optimality of the solutions. More precisely two purposes are achieved with `HamPath`:

- *Shooting equations*: to compute periodic trajectories for the Purcell swimmer, we consider the true Hamiltonian H given by the Pontryagin maximum principle and the associated transversality conditions associated. The normal and regular minimizing curves are the projection of extremals solutions of the following boundary value problem:

$$
\begin{cases}
\dot{q} = \frac{\partial H}{\partial p}, \quad \dot{p} = -\frac{\partial H}{\partial q}, \\
x(0) = x_0, \quad x(T) = x_f, \quad y(0) = y_0, \quad y(T) = y_f \\
\theta_i(T) = \theta_i(0), \ i = 1, 2 \quad \alpha(T) = \alpha(0), \\
p_{\theta_i}(T) = p_{\theta_i}(0), \ i = 1, 2 \quad p_\alpha(T) = p_\alpha(0)
\end{cases}
\tag{2.30}
$$

where $q = (\theta_1, \theta_2, x, y, \alpha)$, $p = (p_{\theta_1}, p_{\theta_2}, p_x, p_y, p_\theta)$ and the final time $T > 0$ is fixed. Due to the sensitivity of the initialization of the shooting algorithm, the latter is initialized with direct methods namely the Bocop toolbox.

- *Local optimality*: to show that the calculated normal stroke is optimal, we perform a rank test on the subspaces spanned by the solutions of the variational equation with suitable initial conditions [21].

Using proposition 14, in the normal case it allows us to check the necessary optimality condition related to the concept of conjugate point. The same holds in the abnormal case using [21].

2.7.1 Nilpotent Approximation

Notations. The state variables are given by $\hat{x} = (\hat{x}_1, \hat{x}_2, \hat{x}_3, \hat{x}_4, \hat{x}_5)$, the adjoint by $\hat{p} = (\hat{p}_1, \hat{p}_2, \hat{p}_3, \hat{p}_4, \hat{p}_5)$, and $\hat{F}_1, \hat{F}_2$ are the vector fields of the normal form given by (2.19). The Hamiltonian lifts are respectively denoted H_1 and H_2.

Normal Case

In the normal case, we consider the extremal system given by the true Hamiltonian described in (2.20). We compute the optimal trajectories with HamPath, and we display on Fig. 2.4 the state and adjoint variables as functions of time. We also illustrate the conjugate points computed according to the algorithm in [27], as well as the smallest singular value for the rank test.

Property on the first conjugate point. Let us consider the fixed energy level $(H_1^2 + H_2^2)_{|t=0} = 1$ along the extremals and the initial state $x(0) = 0$. We take a large number of random initial adjoint vectors $p(0)$ and numerically integrate the extremal system. For each normal extremal, we compute the first conjugate time t_{1c}, the pulsation $\omega = (p_{40}^2 + 4 p_{50}^2)^{1/4}$, and the complete elliptic integral $K(k)$, where k is the amplitude

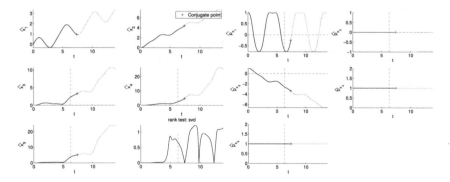

Fig. 2.4 Nilpotent approximation (normal case): state, adjoint variables and first conjugate point (blue cross), with the smallest singular value of the rank test

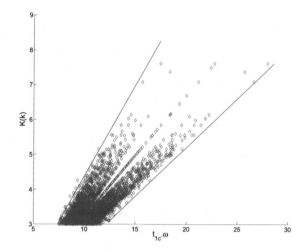

Fig. 2.5 Computations of the complete elliptic integral $K(k, \omega t_c)$ and of the first conjugate point t_{1c} for normal strokes on the energy level $H_1^2 + H_2^2 = 1$. We observe: $0.3\omega t_{1c} - 0.4 < K(k) < 0.5\omega t_{1c} - 0.8$

$$k = \frac{1}{2}\sqrt{\frac{2\sqrt{\hat{p}_{40}^2 + 4\hat{p}_{50}^2} + \hat{p}_{30}^2 - 2\hat{p}_{10}\hat{p}_{40} - 4\hat{p}_{50}\hat{p}_{20} - 4\hat{p}_{50}\hat{p}_{40}\hat{x}_{30}}{\sqrt{\hat{p}_{40}^2 + 4\hat{p}_{50}^2}}}.$$

Let $\gamma(.)$ be a normal extremal starting at $t = 0$ from the origin and defined on $[0, +\infty[$. As illustrated on Fig. 2.5, there exists a first conjugate point along γ corresponding to a conjugate time t_{1c} satisfying the inequality:

$$0.3\omega t_{1c} - 0.4 < K(k) < 0.5\omega t_{1c} - 0.8.$$

Remark 2.3 In Sect. 2.6.4 $u = \omega t + \varphi_0$ is the normalized parametrization of the solutions.

Abnormal Case

Figure 2.6 illustrates the time evolution of the state variables for an abnormal extremal. We check the second order optimality conditions with the algorithm described in [21]. The determinant test and the smallest singular value for the rank condition both indicate that there is no conjugate time for abnormal extremals (Fig. 2.7).

2.7.2 True Mechanical System

We now consider the optimal control problem (2.29) consisting in minimizing either the mechanical energy (2.14) or the criterion $|u|^2$.

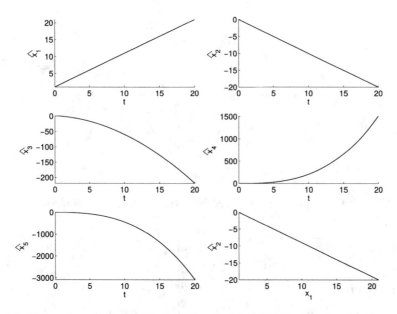

Fig. 2.6   Abnormal case: state variables for $\hat{x}(0) = (1, 0, 1, 0, 0)$, $\hat{p}(0) = (0, 0, -2, 1, 1)$

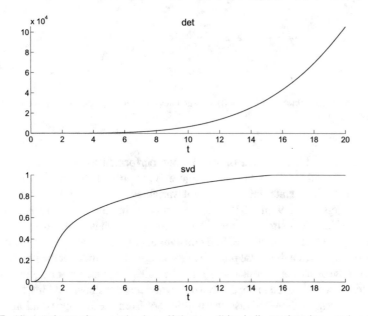

Fig. 2.7   Abnormal case: the second order sufficient condition indicates there is no conjugate point

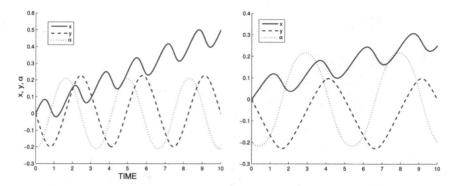

Fig. 2.8 Optimal trajectory for $|u|^2$ (left) and the energy criterion (right)—displayed are the state variables x, y, α

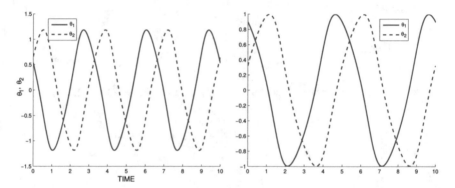

Fig. 2.9 Optimal trajectory for $|u|^2$ (left) and the energy criterion (right)—displayed are the state variables θ_1, θ_2

Direct method. In the first set of simulations performed by Bocop, we set $T = 10$, $x_f = 0.5$, and the bounds $a = 3$ large enough so that the solution is actually unconstrained. The state and the control variables for the optimal trajectory are shown on Figs. 2.8, 2.9 and 2.10, and we observe that the trajectory is actually a sequence of identical strokes. Figure 2.11 displays the phase portrait for the shape angles θ_1, θ_2, which is an ellipse. The constant energy level satisfied by the optimal trajectory implies that the phase portrait of the controls is a circle for the $|u|^2$ criterion, but not for the energy criterion. The adjoint variables (or more accurately in this case, the multipliers associated to the discretized dynamics) are shown on Figs. 2.12, 2.13.

Indirect method. Now we use the multipliers from the Bocop solutions to initialize the shooting algorithm of HamPath. Figures 2.14, 2.15 and 2.16 represent respectively non intersecting strokes and an eight shape stroke. We check the second order optimality conditions according to [27] and observe that there is no conjugate point on $[0, 2\pi]$ for the non intersecting case while a conjugate point is found on $[0, 2\pi]$ for the eight shape stroke.

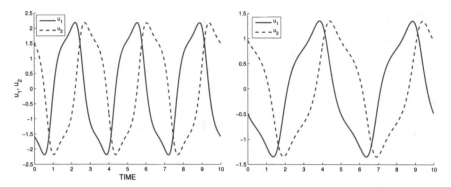

Fig. 2.10 Optimal trajectory for $|u|^2$ (left) and energy criterion (right)—displayed are the control variables

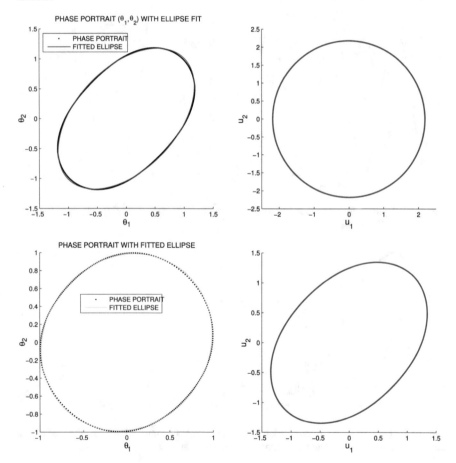

Fig. 2.11 Optimal trajectory for $|u|^2$ (top) and the energy criterion (bottom)—displayed are the phase portrait (ellipse) and the controls

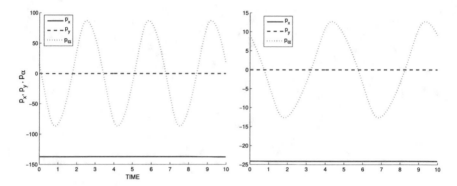

Fig. 2.12 Optimal trajectory for $|u|^2$ (left) and the energy criterion (right)—displayed are the adjoint variables p_x, p_y and p_α

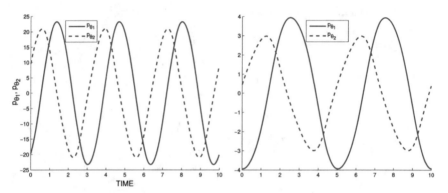

Fig. 2.13 Optimal trajectory for $|u|^2$ (left) and the energy criterion (right)—displayed are the adjoint variables p_{θ_1}, p_{θ_2}

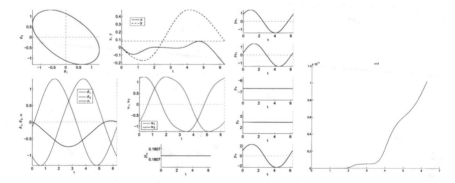

Fig. 2.14 (*Left*) State and adjoint variables for the Purcell swimmer minimizing the mechanical cost. (*Right*) Test of conjugate points (no conjugate point on $[0, 2\pi]$)

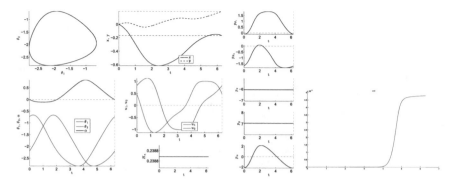

Fig. 2.15 (*Left*) State and adjoint variables for the Purcell swimmer minimizing the mechanical cost. (*Right*) Test of conjugate points (no conjugate point on $[0, 2\pi]$)

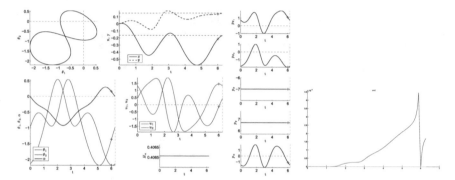

Fig. 2.16 (*Left*) State and adjoint variables for the Purcell swimmer minimizing the mechanical cost. (*Right*) Test of conjugate points. The cross on the trajectories on the left indicates the location of the first conjugate point

Continuation Method

Finally, we construct for the Purcell swimmer, a one parameter family of simple loops strokes using continuation methods (Fig. 2.17).

For the Purcell swimmer, the two families presented in Fig. 2.18 are compared in Fig. 2.19 using the efficiency concept defined as

$$\mathscr{E}(\gamma(\cdot)) = \sqrt{x(T)^2 + y(T)^2}/l(\gamma(\cdot))$$

where $l(\gamma(\cdot))$ is the length of the stroke.

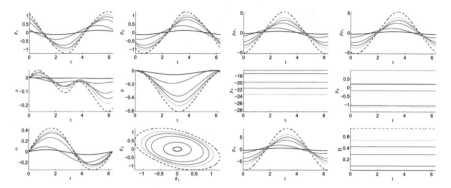

Fig. 2.17 Continuation on the amplitude: $x(T)^2 + y(T)^2 = c_1$ for the $\int_0^T (u_1^2 + u_2^2)\,dt$ cost

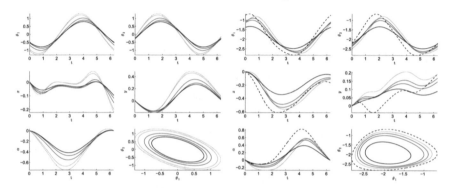

Fig. 2.18 Two families of strokes for the mechanical cost obtained by continuation from the $\int_0^T (u_1^2 + u_2^2)\,dt$ cost to the mechanical cost

2.7.3   Copepod Swimmer

Geometric Analysis of a Copepod Swimmer

In [87], two types of geometric motions are described.

First case: (Fig. 2.20 *(left)*) The two legs are assumed to oscillate sinusoidally according to

$$\theta_1 = \Phi_1 + a\cos(t), \quad \theta_2 = \Phi_2 + a\cos(t + k_2)$$

with $a = \pi/4$, $\Phi_1 = \pi/4$, $\Phi_2 = 3\pi/4$ and $k_2 = \pi/2$. This produces a displacement $x_0(2\pi) = 0.2$.

Second case: (Fig. 2.20 *(right)*) The two legs are paddling in sequence followed by a recovery stroke performed in unison. In this case the controls $u_1 = \dot{\theta}_1$, $u_2 = \dot{\theta}_2$ produce bang arcs to steer the angles between the boundary $\theta_i = 0$ of the domain to the boundary $\theta_i = \pi$, while the unison sequence corresponds to a displacement from π to 0 with the constraint $\theta_1 = \theta_2$.

Our first objective is to relate these properties to geometric optimal control.

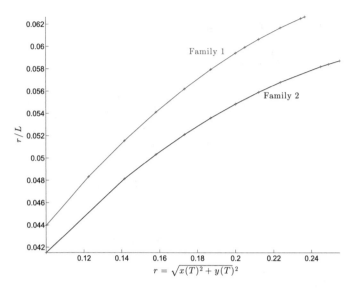

Fig. 2.19 Efficiency curves for the two families of strokes presented in Fig. 2.18

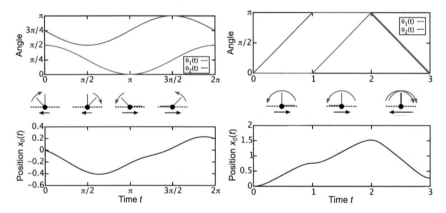

Fig. 2.20 Different geometric motions of the Copepod swimmer. *(left)* Two legs oscillating sinu-soidally according to $\theta_1 = \Phi_1 + a\cos t$ and $\theta_2 = \Phi_2 + a\cos(t + \pi/2)$, where $a = \pi/4$ is the ampli-tude and (Φ_1, Φ_2) is fixed. The displacement after one cycle is $x_0(2\pi) = 0.2$. *(right)* Two legs paddling in sequence. The legs perform power strokes in sequence and then a recovery stroke in unison, each stroke sweeping an angle π

Abnormal Curves in the Copepod Swimmer

Let $q^{\mathsf{T}} = (x_0, \theta_1, \theta_2)$, then the system takes the form:

$$\dot{q}(t) = \sum_{i=1}^{2} u_i(t) F_i(q(t))$$

where the control vector fields are given by:

$$F_i = \frac{\sin \theta_i}{\Delta} \frac{\partial}{\partial x_0} + \frac{\partial}{\partial \theta_i}, \qquad \Delta = 2 + \sin^2 \theta_1 + \sin^2 \theta_2.$$

The Lie brackets in the copepod case are easily calculated and are given by:

$$F_3 = [F_1, F_2] = f(\theta_1, \theta_2)\frac{\partial}{\partial x_0} \text{ with } f(\theta_1, \theta_2) = \frac{2 \sin \theta_1 \sin \theta_2 (\cos \theta_1 - \cos \theta_2)}{\Delta^2},$$

$$[[F_1, F_2], F_1] = \frac{\partial f}{\partial \theta_1}(\theta_1, \theta_2)\frac{\partial}{\partial x_0}, \quad [[F_1, F_2], F_2] = \frac{\partial f}{\partial \theta_2}(\theta_1, \theta_2)\frac{\partial}{\partial x_0}.$$

Lemma 3 *The singular set* $\Sigma : \{q; \det(F_1(q), F_2(q), [F_1, F_2](q)) = 0\}$, *where the vector fields* $F_1, F_2, [F_1, F_2]$ *are coplanar, is given by* $2 \sin \theta_1 \sin \theta_2 (\cos \theta_1 - \cos \theta_2) = 0$ *which is equivalent to:*

- $\theta_i = 0$ *or* π $i = 1, 2$,
- $\theta_1 = \theta_2$

and corresponds to the boundary of the physical domain: $\theta_i \in [0, \pi]$, $\theta_1 \leq \theta_2$, *with respective controls* $u_1 = 0$, $u_2 = 0$ *or* $u_1 = u_2$ *forming a stroke of triangular shape in the phase portait of the variables* θ_1, θ_2.

Remark 2.4 Each point of the boundary is a Martinet point except at the non smooth points (vertices).

The previous lemma provides the interpretation of the triangle shape stroke in terms of abnormal curves.

To understand smooth stroke strategies via optimal control we must introduce the cost function related to the mechanical energy. Recall that according to [76] the mechanical energy of the copepod swimmer is given by:

$$\int_0^T \dot{q}^t M \dot{q} \, dt$$

where $q = (x_0, \theta_1, \theta_2)$ and M is the symmetric matrix:

$$M = \begin{pmatrix} 2 - 1/2(\cos^2(\theta_1) + \cos^2(\theta_2)) & -1/2 \sin(\theta_1) & -1/2 \sin(\theta_2) \\ -1/2 \sin(\theta_1) & 1/3 & 0 \\ -1/2 \sin(\theta_2) & 0 & 1/3 \end{pmatrix}. \qquad (2.31)$$

Taking into account the constraints on the velocities, the integrand can be written as:

$$a(q)u_1^2 + 2b(q)u_1 u_2 + c(q)u_2^2$$

where

$$a = \frac{1}{3} - \frac{\sin^2 \theta_1}{2(2 + \sin^2 \theta_1 + \sin^2 \theta_2)}, \quad b = -\frac{\sin \theta_1 \sin \theta_2}{2(2 + \sin^2 \theta_1 + \sin^2 \theta_2)},$$

$$c = \frac{1}{3} - \frac{\sin^2 \theta_2}{2(2 + \sin^2 \theta_1 + \sin^2 \theta_2)}.$$

The pseudo-Hamiltonian is then expressed as:

$$H(q, p, p^0) = u_1 H_1(q, p) + u_2 H_2(q, p) + p^0 \left(a(q)u_1^2 + 2b(q)u_1u_2 + c(q)u_2^2 \right).$$

Taking $p^0 = -1/2$, the normal controls are computed by solving the equations:

$$\frac{\partial H}{\partial u_1} = 0, \quad \frac{\partial H}{\partial u_2} = 0.$$

We obtain:

$$u_1 = -\frac{3(4H_1 + 2H_1 \sin^2 \theta_1 + 3H_2 \sin \theta_1 \sin \theta_2 - H_1 \sin^2 \theta_2)}{\sin^2 \theta_1 + \sin^2 \theta_2 - 4},$$

$$u_2 = -\frac{9H_1 \sin \theta_1 \sin \theta_2 + 6H_2(2 + \sin^2 \theta_2) - 3H_2 \sin^2 \theta_1}{\sin^2 \theta_1 + \sin^2 \theta_2 - 4}.$$

and plugging this control u back into the pseudo-Hamiltonian provides the true Hamiltonian which we denote by H_n.

Note also that H_n can also be obtained by constructing an orthonormal basis of the metric using a feedback transformation $u = \beta(q)v$ to transform the problem into:

$$\dot{q} = (F\beta(q))(v), \quad \min_{v(\cdot)} \int_0^T (v_1^2(t) + v_2^2(t)) \, dt$$

where F is the matrix (F_1, F_2). Writing $F\beta = (F_1', F_2')$, F_1', F_2' will form an orthonormal frame. The computation is straightforward and the normal Hamiltonian H_n takes the form $H_n = \frac{1}{2}(H_1'^2 + H_2'^2)$ where H_i' is the Hamiltonian lift of F_i'.

The Concept of Efficiency

To compare strokes with different amplitudes we introduce the following definition of efficiency [69].

Definition 37 The efficiency of a stroke $\gamma(\cdot)$ is defined by:

$$E(\gamma(\cdot)) = x_0(T)/L(\gamma(\cdot))$$

where x_0 is the displacement of the swimmer and L is the length of the curve $\gamma(.)$.

Fig. 2.21 Closed periodic
planar curves: non
intersecting curve, eight
curve and limaçon curve

The transversality condition given in Exercise 1.1 can be generalized, see [89]. For instance, for the copepod swimmer, considering the augmented adjoint vector (p, p^0), the transversality condition implies that:

$(p(T), p^0(T))$ is collinear to the gradient of the set $E = c$, where c is a constant.

Geometric Classification of Smooth Strokes

The expected strokes are related to the classification of smooth periodic curves in the plane up to a diffeomorphism, assuming that in our discussion we relax the state constraints on the shape variable. This problem was studied by Whitney (1937) and Arnold (1994), see [14]. In this classification we have in particular the three cases of Fig. 2.21.

Each of this curve has a specific physical interpretation for the swimmer problem.

Numerical Computations

- *Micro-local analysis.* First, we compute the normal strokes using the Maximum Principle to recover the strokes displayed in Fig. 2.21. Below, we present the numerical calculations of these strokes using the weak Maximum Principle.

  An important point is to account for the transversality conditions associated with the periodicity requirement $\theta_i(0) = \theta_i(2\pi)$, $i = 1, 2$ which are given by:

  $$p_{\theta_i}(0) = p_{\theta_i}(2\pi), \ i = 1, 2.$$

  The solutions are computed via a shooting method using the HamPath code. Finally, we evaluate numerically the value function which reduces to $2\pi H_n$ the given reference geodesic, since H_n is constant.
- *Second order optimality.* Conjugate points are computed for each type of stroke which leads to select simple loops as candidates for minimizers, see Fig. 2.22.
- *Abnormal triangle.* To deal with the global optimality problem we use the *geometric efficiency* $E = x_0/L$ for single loops constrained in the triangle (see Fig. 2.24 and Table 2.1). From our analysis we deduce that the (triangle) abnormal stroke is not optimal. Indeed, one can choose a normal stroke (inside the triangle) such that the displacement is $\bar{x}_0/2$ with $\bar{x}_0 = 2.742$ and length $< \bar{L}/2$ where $\bar{L}$ =length of the triangle. Applying twice the normal stroke, we obtain the same displacement $\bar{x}_0$ than with the abnormal stroke but with a length $< \bar{L}$.

Therefore, we proved the following theorem.

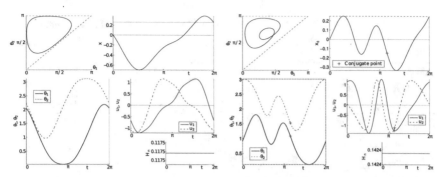

Fig. 2.22 *(Left)* Normal stroke where the constraints are satisfied: simple loop with no conjugate point on $[0, T]$. *(Right)* Limaçon with inner loop with one conjugate point on $[0, T]$.

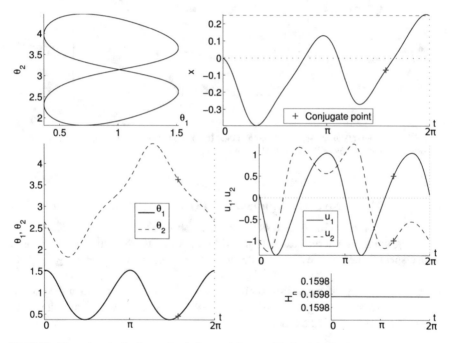

Fig. 2.23 Normal stroke for the mechanical cost: eight case. We fixed the displacement to $x_0(2\pi) = 0.25$

Theorem 8 *The abnormal triangle is not optimal for both costs: minimizing length with fixed displacement or maximizing the efficiency (Fig. 2.23).*

So far, the copepod microswimmer was analyzed using mainly simulations but a complete analysis can be obtained combining mathematical analysis based on numerical evidence. We proceed as follows.

First, to simplify the computations and to have a clear interpretation of the pictures in the Euclidean frame, we replace the mechanical energy by the Euclidean

Table 2.1 Ratio x_0/L for the abnormal stroke and different normal strokes corresponding to the mechanical cost

Types of γ	x_0	$L(\gamma)$	$x_0/L(\gamma)$
Abnormal	2.742e-1	4.933	5.558e-2
Simple loop (Fig. 2.22, left)	2.600e-1	3.046	**8.536e-2**
Limaçon (Fig. 2.22, right)	2.500e-1	3.353	7.456e-2
Simple loop with small amplitude	0.500e-1	9.935e-1	5.033e-2

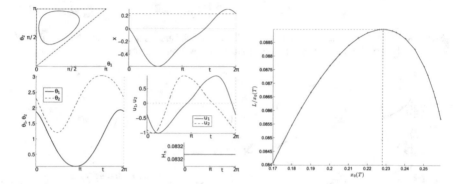

Fig. 2.24 Efficiency curve for the mechanical cost (top) and the corresponding maximizing curve (bottom). The efficiency of the abnormal curve is $5.56e^{-2}$

cost $\int_0^T |u(t)|^2 \, dt$. Note that the true cost case can be analyzed using a numerical continuation between the two costs (HamPath software).

Using the nilpotent approximation and Lemma 3, one must consider two cases with respect to the triangle $\mathscr{T}$ associated with the state constraints: $0 \leq \theta_1 \leq \theta_2 \leq \pi$.

Point interior to the triangle. Take such a point $q = (x_0, \theta_1, \theta_2)$. Then near the chosen point, there exists privileged coordinates $\hat{x} = (\hat{x}_1, \hat{x}_2, \hat{x}_3)$ such that the nilpotent SR-model is given by the Dido model:

$$\dot{\hat{x}} = u_1 \hat{F}_1(\hat{x}) + u_2 \hat{F}_2(\hat{x}), \qquad \min_u \int_0^T (u_1(t)^2 + u_2^2(t)) \, dt$$

with

$$\hat{F}_1 = \frac{\partial}{\partial \hat{x}_1} + \hat{x}_2 \frac{\partial}{\partial \hat{x}_3}, \qquad \hat{F}_2 = \frac{\partial}{\partial \hat{x}_2} - \hat{x}_1 \frac{\partial}{\partial \hat{x}_3}.$$

This model implies that starting from each q we have a one parameter family of symmetric simple strokes (see Fig. 2.25)

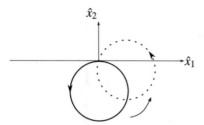

Fig. 2.25 One parameter family of circles which are the geodesics of the Heisenberg-Brockett problem

Points on the sides of the triangle but different of the vertices. Take such a point $q = (x_0, \theta_1, \theta_2)$. Then the SR-nilpotent model is the Martinet flat case. Thus, one can find privileged coordinates $\hat{x} = (\hat{x}_1, \hat{x}_2, \hat{x}_3)$ such that the model is:

$$\dot{\hat{x}} = u_1 \hat{F}_1(\hat{x}) + u_2 \hat{F}_2(\hat{x}), \qquad \min_u \int_0^T (u_1^2(t) + u_2^2(t))\, dt$$

where

$$\hat{F}_1 = \frac{\partial}{\partial \hat{x}_1} + \frac{\hat{x}_2^2}{2} \frac{\partial}{\partial \hat{x}_3}, \quad \hat{F}_2 = \frac{\partial}{\partial \hat{x}_2}.$$

This model leads to the calculation of eight strokes parameterized by elliptic functions which correspond to lemniscates of Bernoulli.

All these models are not stable models and higher order approximations can be used to generate strokes with small amplitudes. Also by perturbation at a interior point of the triangle, we can obtain limaçon's strokes by doubling the period. This is indeed confirmed by numerical simulations using the true model and represented on Fig. 2.26.

Moreover for the true system with the Euclidean cost, the numerical simulations show the existence of a one-parameter family of simple strokes symmetric with respect to the axis $\mathscr{D} : \theta_2 = -\theta_1 + \pi$. They are obtained by integrating from $\mathscr{D}$ identified to a cross-section and with a tangent vector taken normal to $\mathscr{D}$, each stroke being associated with a different energy level, see Fig. 2.27.

It leads to the following proposition.

Proposition 15 *There exists a one parameter family of simple strokes, symmetric with respect to the $\mathscr{D}$-axis and foliating the interior of the triangle $\mathscr{T}$, each associated to a different energy level.*

The final result of our analysis is captured in the following theorem.

Theorem 9 *Among this family of strokes, there exists a unique stroke with maximal efficiency among all the strokes of the copepod swimmer for the $\int_0^T (u_1^2(t) + u_2^2(t))\, dt$ cost.*

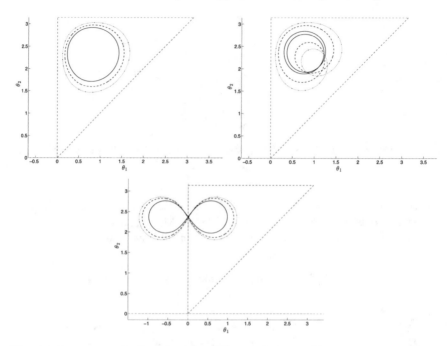

Fig. 2.26 One parameter family of simple loops, limaçons and Bernoulli lemniscates normal strokes

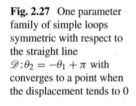

Fig. 2.27 One parameter
family of simple loops
symmetric with respect to
the straight line
$\mathscr{D}: \theta_2 = -\theta_1 + \pi$ with
converges to a point when
the displacement tends to 0

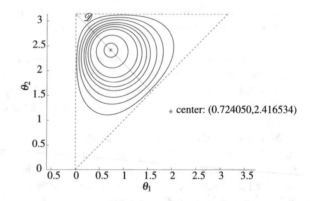

$*$ center: $(0.724050, 2.416534)$

Sketch of the Proof

First we have the following lemma.

Lemma 2.1 *For the Euclidean case (or the mechanical energy case) the geodesic flow is invariant under the transformation* $\delta : (\theta_1, \theta_2, x_0) \mapsto (\pi - \theta_2, \pi - \theta_1, x_0)$.

From this, we deduce that the one parameter family of simple loops represented on Fig. 2.27 is symmetric with respect to the straight line $\mathscr{D}$. The center of this family can be calculated as follows. We choose a point $\theta(0) = (\theta_1(0), \theta_2(0))$ on the line $\mathscr{D}$ which can be identified to $(0, 0)$ if we introduce the new coordinates $x =$

$\theta_1 - \theta_1(0)$, $y = \theta_2 - \theta_2(0)$. Using a transformation of the form $Z = z - c_1 x - c_2 y$ we get a graded set of coordinates (x, y, Z) with weights $(1, 1, 2)$ establishing a link between the physical coordinates $(\theta_1, \theta_2, x_0)$ and the privileged coordinates identified to (x, y, Z). Using this gradation, the nilpotent (order -1) SR-model is given by the Dido model. This model is not stable under perturbation and higher order terms have to be taken into account. In particular, using the weights $(1, 1, 2)$ the model of order 0 can be computed. Using the analysis of [31], the model of order zero can be identified with the model of order -1 using diffeomorphism and feedback preserving the Euclidean energy. A precise computation detailed in [16] shows that the only point $\theta(0)$ such that the diffeomorphism is not mixing the shape variable θ with the displacement variable x_0 corresponds to the center $\theta(0) \simeq (0.72, 2.41)$ of Fig. 2.27. Hence, we proved that there exists only one point to generate such a family of simple loops (compare with Fig. 2.18 in the Purcell case).

Now, we must prove that the only strokes candidates as minimizers in the interior of the triangle are simple strokes. This can be proved using the Stokes theorem and the following lemma.

Lemma 2.2 *Consider the smooth one-form on $\mathbb{R}^2$: $\omega := \sum_{i=1}^{2} \dfrac{\sin \theta_i}{\Delta(\theta)} d\theta_i$ with $\Delta(\theta) = 2 + \sin^2 \theta_1 + \sin^2 \theta_2$ and introduce $f(\theta) = 2 \sin \theta_1 \sin \theta_2 (\cos \theta_1 - \cos \theta_2)/\Delta(\theta)^2$. Then,*

1. $d\omega = -f(\theta_1, \theta_2)d\theta_1 \wedge d\theta_2$.
2. For any bounded Stokes domain $D \subset \mathbb{R}^2$, we have

$$\oint_{\partial D} \omega = \int_D d\omega$$

and if γ is a piecewise smooth stroke with $\gamma = \partial D$ the associated displacement is

$$x_0(T) = \int_D d\omega.$$

3. $d\omega < 0$ in the interior of the triangle $\mathcal{T} : 0 \le \theta_1 \le \theta_2 \le \pi$, and $d\omega$ vanishes on the boundary of $\mathcal{T}$ formed by the abnormal stroke.

In particular this lemma allows to compare efficiency of simple loops versus limaçons and eight shape strokes in the interior of the triangle.

Another method from optimal control theory is to compute conjugate points. This can be performed by numerical computations but more theoretical computations are related to conjugate loci computations on the SR-sphere. In particular, for limaçons with small amplitudes, conjugate points can be estimated as follows. According to the Dido model, the only strokes with small amplitudes can be either simple loops or limaçons, obtained by perturbation of a simple loop followed twice. For the Dido model, using the explicit computation, the first conjugate point appears on a simple loop after exactly one period. By perturbation, for a simple stroke with small amplitude, the first conjugate time corresponds approximately to the period. Hence a limaçon of small amplitude produced by period doubling has necessarily a conjugate point. This gives a rigorous proof of the existence of conjugate point for limaçons with small amplitude.

2.8   Conclusion and Bibliographic Remarks

We made a short presentation of the problem of microswimming using the Purcell and the copepod case in the frame of SR-geometry, combining analytic and numeric methods in optimal control based on the analysis of the geodesic flow to determine the most efficient stroke. A different approach combining Stokes theorem to determine the shape of optimal strokes and direct numeric methods using Fourier analysis were used earlier in a series of articles, see for instance [10].

Note also that the copepod case is the analog of a limit of symmetric Purcell swimmer described and analyzed in [10].

The approaches are complementary. The main result of this theory is the existence of center of swimmings from which are emanating a one parameter family of simple strokes to compute the most efficient stroke. See [6] for an earlier computation using a shooting method.

Note also the (geometric) link of microswimmers in SR-geometry with the geodesic motion of a 2D-particle in a magnetic field very well presented in [74]. This leads to a fine and technical study in [2] as a generalization of the Dido problem, to compute conjugate and cut loci for small lengths. Such results being applicable to generate in general conjugate and cut loci, using numeric continuation methods.

Chapter 3
Maximum Principle and Application to Nuclear Magnetic Resonance and Magnetic Resonance Imaging

3.1 Maximum Principle

In this section we state the Pontryagin maximum principle and we outline the proof. We adopt the presentation from Lee and Markus [64] where the result is presented into two theorems. The complete proof is complicated but rather standard, see the original book from the authors [77].

Theorem 10 *We consider a system of* $\mathbb{R}^n$: $\dot{x}(t) = f(x(t), u(t))$, *where* $f : \mathbb{R}^{n+m} \to \mathbb{R}^n$ *is a* C^1*-mapping. The family* $\mathcal{U}$ *of admissible controls is the set of bounded measurable mappings* $u(\cdot)$, *defined on* $[0, T]$ *with values in a control domain* $\Omega \subset \mathbb{R}^m$ *such that the response* $x(\cdot, x_0, u)$ *is defined on* $[0, T]$. *Let* $\bar{u}(\cdot) \in \mathcal{U}$ *be a control and let* $\bar{x}(\cdot)$ *be the associated trajectory such that* $\bar{x}(T)$ *belongs to the boundary of the accessibility set* $A(x_0, T)$. *Then there exists* $\bar{p}(\cdot) \in \mathbb{R}^n \setminus \{0\}$, *an absolutely continuous function defined on* $[0, T]$ *solution almost everywhere of the adjoint system:*

$$\dot{p}(t) = -p(t)\frac{\partial f}{\partial x}(\bar{x}(t), \bar{u}(t)) \tag{3.1}$$

such that for almost every $t \in [0, T]$ *we have*

$$H(\bar{x}(t), \bar{p}(t), \bar{u}(t)) = M(\bar{x}(t), \bar{p}(t)) \tag{3.2}$$

where

$$H(x, p, u) = \langle p, f(x, u) \rangle$$

© The Author(s), under exclusive license to Springer International Publishing AG, part of Springer Nature 2018
B. Bonnard et al., *Geometric and Numerical Optimal Control*, SpringerBriefs in Mathematics, https://doi.org/10.1007/978-3-319-94791-4_3

and

$$M(x, p) = \max_{u \in \Omega} H(x, p, u).$$

Moreover $t \mapsto M(\bar{x}(t), \bar{p}(t))$ is constant on $[0, T]$.

Proof The accessibility set is not in general convex and it must be approximated along the reference trajectory $\bar{x}(\cdot)$ by a convex cone. The approximation is obtained by using *needle type variations* of the control $\bar{u}(\cdot)$ which are closed for the L^1-topology. (We do not use L^∞ perturbations and the Fréchet derivative of the end-point mapping computed in this Banach space.)

Needle Type Approximation
We say that $0 \leq t_1 \leq T$ is a *regular time* for the reference trajectory if

$$\frac{d}{dt}\Big|_{t=t_1} \int_0^t f(\bar{x}(\tau), \bar{u}(\tau)) d\tau = f(\bar{x}(t_1), \bar{u}(t_1))$$

and from measure theory we have that almost every point of $[0, T]$ is regular.

At a regular time t_1, we define the following L^1-perturbation $\bar{u}_\varepsilon(\cdot)$ of the reference control: we fix $l, \varepsilon \geq 0$ small enough and we set

$$\bar{u}_\varepsilon(t) = \begin{cases} u_1 \in \Omega & \text{constant} \quad \text{on } [t_1 - l\varepsilon, t_1] \\ \bar{u}(t) & \text{otherwise on } [0, T] \end{cases}.$$

We denote by $\bar{x}_\varepsilon(\cdot)$ the associated trajectory starting at $\bar{x}_\varepsilon(0) = x_0$. We denote by $\varepsilon \mapsto \alpha_t(\varepsilon)$ the curve defined by $\alpha_t(\varepsilon) = \bar{x}_\varepsilon(t)$ for $t \geq t_1$. We have

$$\bar{x}_\varepsilon(t_1) = \bar{x}(t_1 - l\varepsilon) + \int_{t_1 - l\varepsilon}^{t_1} f(\bar{x}_\varepsilon(t), \bar{u}_\varepsilon(t)) dt$$

where $\bar{u}_\varepsilon = u_1$ on $[t_1 - l\varepsilon, t_1]$, Moreover

$$\bar{x}(t_1) = \bar{x}(t_1 - l\varepsilon) + \int_{t_1 - l\varepsilon}^{t_1} f(\bar{x}(t), \bar{u}(t)) dt$$

and since t_1 is a regular time for $\bar{x}(\cdot)$ we have

$$\bar{x}_\varepsilon(t_1) - \bar{x}(t_1) = l\varepsilon(f(\bar{x}(t_1), u_1) - f(\bar{x}(t_1), \bar{u}(t_1))) + o(\varepsilon).$$

In particular if we consider the curve $\varepsilon \mapsto \alpha_{t_1}(\varepsilon)$, it is a curve with origin $\bar{x}(t_1)$ and whose tangent vector is given by

$$v = l(f(\bar{x}(t_1), u_1) - f(\bar{x}(t_1), \bar{u}(t_1))). \tag{3.3}$$

For $t \geq t_1$, consider the local diffeomorphism: $\phi_t(y) = x(t, t_1, y, \bar{u})$ where $x(\cdot, t_1, y, \bar{u})$ is the solution corresponding to $\bar{u}(\cdot)$ and starting at $t = t_1$ from y. By construction we have $\alpha_t(\varepsilon) = \phi_t(\alpha_t(\varepsilon))$ for ε small enough and moreover for $t \geq t_1$, $v_t = \frac{d}{d\varepsilon}\big|_{\varepsilon=0} \alpha_t(\varepsilon)$ is the image of v by the Jacobian $\frac{\partial \phi_t}{\partial x}$. In other words v_t is the solution at time t of the variated equation

$$\frac{dV}{dt} = \frac{\partial f}{\partial x}(\bar{x}(t), \bar{u}(t))V \tag{3.4}$$

with condition $v_t = v$ for $t = t_1$. We can extend v_t on the whole interval $[0, T]$. The construction can be done for an arbitrary choice of t_1, l and u_1. Let $\Pi = \{t, l, u_1\}$ be fixed, we denote by $v_\Pi(t)$ the corresponding vector v_t.

Additivity Property

Let t_1, t_2 be two regular points of $\bar{u}(\cdot)$ with $t_1 < t_2$ and l_1, l_2 small enough. We define the following perturbation

$$\bar{u}_\varepsilon(t) = \begin{cases} u_1 \text{ on } [t_1 - l_1\varepsilon, t_1] \\ u_2 \text{ on } [t_2 - l_2\varepsilon, t_2] \\ \bar{u}(t); \text{ otherwise on } [0, T] \end{cases}$$

where u_1, u_2 are constant values of Ω and let $\bar{x}_\varepsilon(\cdot)$ be the corresponding trajectory. Using the composition of the two elementary perturbations $\Pi_1 = \{t_1, l_1, u_1\}$ and $\Pi_2 = \{t_2, l_2, u_2\}$ we define a new perturbation $\Pi : \{t_1, t_2, l_1, l_2, u_1, u_2\}$. If we denote by $v_{\Pi_1}(t), v_{\Pi_2}(t)$ and $v_\Pi(t)$ the respective tangent vectors, a computation similar to the previous one gives us:

$$v_\Pi(t) = v_{\Pi_1}(t) + v_{\Pi_2}(t), \qquad \text{for } t \geq t_2.$$

We can deduce the following lemma.

Lemma 4 *Let $\Pi = \{t_1, \ldots, t_s, \lambda_1 l_1, \ldots, \lambda_s l_s, u_1, \ldots, u_s\}$ be a perturbation at regular times $t_i, t_1 < \cdots < t_s, l_i \geq 0, \lambda_i \geq 0, \sum_{i=1}^s \lambda_i = 1$ and corresponding to elementary perturbations $\Pi_i = \{t_i, l_i, u_i\}$ with tangent vectors $v_{\Pi_i}(t)$. Let $\bar{x}_\varepsilon(\cdot)$ be the associated response with perturbation Π. Then we have*

$$\bar{x}_\varepsilon(t) = \bar{x}(t) + \sum_{i=1}^s \varepsilon\lambda_i v_{\Pi_i}(t) + o(\varepsilon) \tag{3.5}$$

where $\frac{o(\varepsilon)}{\varepsilon} \to 0$, uniformly for $0 \leq t \leq T$ and $0 \leq \lambda_i \leq 1$.

Definition 38 Let $\bar{u}(\cdot)$ be an admissible control and $\bar{x}(\cdot)$ its associated trajectory defined for $0 \leq t \leq T$. The first Pontryagin's cone $K(t), 0 < t \leq T$ is the smallest convex cone at $\bar{x}(t)$ containing all elementary perturbation vectors for all regular times t_i.

Definition 39 Let $v_1, \ldots, v_n$ be linearly independent vectors of $K(t)$, each v_i being formed as convex combinations of elementary perturbation vectors at distinct times. An elementary simplex cone C is the convex hull of the vectors v_i.

Lemma 5 *Let v be a vector interior to $K(t)$. Then there exists an elementary simplex cone C containing v in its interior.*

Proof In the construction of the interior of $K(t)$, we use the convex combination of elementary perturbation vectors at regular times not necessarily distinct. Clearly by continuity we can replace such a combination with n distinct times.

Approximation Lemma
An important technical lemma is the following topological result whose proof uses the Brouwer fixed point theorem.

Lemma 6 *Let v be a nonzero vector interior to $K(t)$, then there exists $\lambda > 0$ and a conic neighborhood N of λv such that N is contained in the accessibility set $A(x_0, T)$.*

Proof See [64].

Separation Step
To finish the proof, we use the geometric Hahn-Banach theorem. Indeed if $\bar{x}(T) \in \partial A(x_0, T)$ there exists a sequence $x_n \notin A(x_0, T)$ such that $x_n \to \bar{x}(T)$ when $n \to +\infty$ and the unit vectors $\frac{x_n - x(T)}{|x_n - x(T)|}$ have a limit ω when $n \to \infty$. The vector ω is not interior to $K(T)$ otherwise from Lemma 6 there would exist $\lambda > 0$ and a conic neighborhood of $\lambda \omega$ in $A(x_0, T)$ and this contradicts the fact that $x_n \notin A(x_0, T)$ for any n. Let π be any hyperplane at $\bar{x}(T)$ separating $K(T)$ from ω and let $\bar{p}$ be the exterior unit normal to π at $\bar{x}(T)$. Let us define $\bar{p}(\cdot)$ as the solution of the adjoint equation

$$\dot{p}(t) = -p(t)\frac{\partial f}{\partial x}(\bar{x}(t), \bar{u}(t))$$

satisfying $p(T) = \bar{p}$. By construction we have

$$\bar{p}(T)v(T) \leq 0$$

for each elementary perturbation vector $v(T) \in K(T)$ and since for $t \in [0, T]$ the following equations hold:

$$\dot{\bar{p}}(t) = -\bar{p}(t)\frac{\partial f}{\partial x}(\bar{x}, \bar{u}), \qquad \dot{v}(t) = \frac{\partial f}{\partial x}(\bar{x}, \bar{u})v$$

we have

$$\frac{d}{dt}\bar{p}(t)v(t) = 0.$$

Hence $\bar{p}(t)v(t) = \bar{p}(T)v(T) \leq 0, \forall t$. Assume that the maximization condition (3.2) is not satisfied on some subset S of $0 \leq t \leq T$ with positive measure. Let $t_1 \in S$ be a regular time, then there exists $u_1 \in \Omega$ such that

$$\bar{p}(t_1)f(\bar{x}(t_1), \bar{u}(t_1)) < \bar{p}(t_1)f(\bar{x}(t_1), u_1).$$

Let us consider the elementary perturbation $\Pi_1 = \{t_1, l, u_1\}$ and its tangent vector

$$v_{\Pi_1}(t_1) = l\left[f(\bar{x}(t_1), u_1) - f(\bar{x}(t_1), \bar{u}(t_1))\right].$$

Then using the above inequality we have that

$$\bar{p}(t_1)v_{\Pi_1}(t_1) > 0$$

which contradicts $\bar{p}(t_1)v_{\Pi_1}(t_1) \leq 0$, for all t. Therefore the inequality

$$H(\bar{x}(t), \bar{p}(t), \bar{u}(t)) = M(\bar{x}(t), \bar{p}(t))$$

is satisfied almost everywhere on $0 \leq t \leq T$. Using a standard reasoning we can prove that $t \mapsto M(\bar{x}(t), \bar{p}(t))$ is absolutely continuous and has zero derivative almost everywhere on $0 \leq t \leq T$, see [64].

Theorem 11 *Let us consider a general control system:* $\dot{x}(t) = f(x(t), u(t))$ *where* f *is a continuously differentiable function and let* M_0, M_1 *be two* C^1 *submanifolds of* $\mathbb{R}^n$. *We assume the set* $\mathcal{U}$ *of admissible controls to be the set of bounded measurable mappings* $u : [0, T(u)] \to \Omega \in \mathbb{R}^m$, *where* Ω *is a given subset of* $\mathbb{R}^m$. *Consider the following minimization problem:* $\min_{u \in \mathcal{U}} C(u)$, $C(u) = \int_0^T f^0(x(t), u(t))dt$ *where* $f^0 \in C^1$, $x(0) \in M_0, x(T) \in M_1$ *and* T *is not fixed. We introduce the augmented system:*

$$\dot{x}^0(t) = f^0(x(t), u(t)), \qquad x^0(0) = 0, \tag{3.6}$$

$$\dot{x}(t) = f(x(t), u(t)), \tag{3.7}$$

$\hat{x}(t) = (x^0(t), x(t)) \in \mathbb{R}^{n+1}$, $\hat{f} = (f^0, f)$. *If* $(x^*(\cdot), u^*(\cdot)$ *is optimal on* $[0, T^*]$, *then there exists* $\hat{p}^*(\cdot) = (p^0, p(\cdot)) : [0, T^*] \to \mathbb{R}^{n+1} \setminus \{0\}$ *absolutely continuous, such that* $(\hat{x}^*(\cdot), \hat{p}^*(\cdot), u^*(\cdot))$ *satisfies the following equations almost everywhere on* $0 \leq t \leq T^*$:

$$\dot{\hat{x}}(t) = \frac{\partial \hat{H}}{\partial \hat{p}}(x(t), \hat{p}(t), u(t)), \quad \dot{\hat{p}}(t) = -\frac{\partial \hat{H}}{\partial \hat{x}}(x(t), \hat{p}(t), u(t)) \tag{3.8}$$

$$\hat{H}(x(t), \hat{p}(t), u(t)) = \hat{M}(x(t), \hat{p}(t)) \tag{3.9}$$

where

$$\hat{H}(x(t), \hat{p}(t), u(t)) = \langle \hat{p}, \hat{f}(x, u) \rangle, \ \hat{M}(\hat{x}, \hat{p}) = \max_{u \in \Omega} \hat{H}(\hat{x}, \hat{p}, u).$$

Moreover, we have

$$\hat{M}(x(t), \hat{p}(t)) = 0, \forall t, \ p^0 \leq 0 \tag{3.10}$$

and the boundary conditions (transversality conditions):

$$x^*(0) \in M_0, \ x^*(T^*) \in M_1, \tag{3.11}$$

$$p^*(0) \perp T_{x^*(0)}M_0, \ p^*(T^*) \perp T_{x^*(T)}M_1. \tag{3.12}$$

Proof (For the complete proof, see [64] or [77]) Since $(x^*(\cdot), u^*(\cdot))$ is optimal on $[0, T^*]$, the augmented trajectory $t \mapsto \hat{x}^*(t)$ is such that $\hat{x}^*(T)$ belongs to the boundary of the accessibility set $\hat{A}(x^*(0), T^*)$. Hence by applying Theorem 10 to the augmented system, one gets the conditions (3.8), (3.9) and $\hat{M}$ constant. To show that $\hat{M} \equiv 0$, we construct an approximated cone $K'(T)$ containing $K(T)$ but also the two vectors $\pm \hat{f}(x^*(T), u^*(T))$ using time variations (the transfer time is not fixed). To prove the transversality conditions, we use a standard separation lemma as in the proof of Theorem 10.

Definition 40 A triplet $(x(\cdot), p(\cdot), u(\cdot))$ solution of the maximum principle is called an extremal.

3.2 Special Cases

Minimal Time
Consider the time minimum case: $f^0 = 1$. In this case, an optimal control u^* on $[0, t^*]$ is such that the corresponding trajectory $x^*(.)$ is such that for each $t > 0, x^*(t)$ belongs to the boundary of the accessibility set $A(x^*(0), t)$. The pseudo-Hamiltonian of the augmented system is written:

$$\hat{H}(\hat{x}, \hat{p}, u) = H(x, p, u) + p_0$$

with $H(x, p, u)$ is the reduced pseudo-Hamiltonian and since $p_0 \leq 0$, conditions 3.9, 3.10 become

$$H(x^*(t), p^*(t), u^*(t)) = M(x^*(t), p^*(t)) \quad a.e.$$

with $M(x, p) = \underset{u \in \Omega}{\text{Max}} H(x, p, u)$ and $M(x^*(t), p^*(t)) \geq 0$ is constant everywhere.

Mayer Problem
A *Mayer problem* is an optimal control problem for a system $\frac{dx}{dt} = f(x, u), \ u \in \Omega, \ x(0) = x_0$, where the cost to be minimized is of the form:

$$\underset{u \in \Omega}{\text{Min}} \ c(x(t_f))$$

where $c : \mathbb{R}^n \longrightarrow \mathbb{R}$ is smooth the transfer time t_f is fixed and the final boundary conditions are of the form: $g(x(t_f))$, with $g : \mathbb{R}^n \to \mathbb{R}^k$ is smooth.

In this case the maximum principle and the geometric interpretation of this principle lead to the following:

- Each optimal control u^* on $[0, t_f]$ with response $x^*(.)$ is such that $x^*(t_f)$ belongs to the boundary of the accessibility set $A(x_0, t_f)$ and at the final point the adjoint vector $p^*(t_f)$ is orthogonal to the manifold defined by the boundary conditions and the cost function:

$$M \; : \; \{\, x; \; g(x) = 0, \; c(x) = m \,\}$$

where m is the minimum.

Introducing the pseudo-Hamiltonian

$$H(x, p, u) = \langle p, f(x, u) \rangle$$

the necessary optimality conditions are:

$$\frac{dx^*}{dt} = \frac{\partial H}{\partial p}(x^*, p^*, u^*), \quad \frac{dp^*}{dt} = -\frac{\partial H}{\partial x}(x^*, p^*, u^*), \tag{3.13}$$

$$H(x^*, p^*, u^*) = \max_{u \in \Omega} H(x^*, p^*, u) \tag{3.14}$$

and the following boundary conditions

$$f(x^*(t_f)) = 0, \tag{3.15}$$

$$p^*(t_f) = p_0.\frac{\partial c}{\partial x}(x^*(t_f)) + \delta.\frac{\partial g}{\partial x}(x^*(t_f)), \tag{3.16}$$

$$p_0 \leq 0 \text{ (transversality conditions)}.$$

Exercise 3.1 Write the necessary optimality conditions for a *Bolza problem* where the cost problem is of the form:

$$C(u) = g(x(t_f)) + \int_0^{t_f} f^0(x(t), u(t))dt.$$

3.3 Application to NMR and MRI

Optimal control was very recently applied very successively to a general research project initiated by S. Glaser: the control of spins systems with applications to Nuclear Magnetic Resonance (NMR) and Magnetic Resonance Imaging (MRI). Such success is partially explained by an accurate representation of the control problem by the *Bloch equations* introduced in 1946 and F. Bloch and E.M. Purcell were awarded the 1952 Nobel Prizes for Physics for "their development of new ways and method for NMR", Purcell providing a nice link between our two working examples.

Next, we make a mathematical introduction of Bloch equations and the concept of *resonance*, in order to model the class of associate problems objects of our research program.

3.3.1 Model

The Bloch equations are a set of macroscopic equations which accurately describe the experimental model in NMR and MRI [66] based on the concept of the dynamics of a spin-1/2 particle. At this level it is represented by a *magnetization vector* $M = (M_x, M_y, M_z)$ in the *laboratory reference frame* which evolves according to

$$\frac{dM}{d\tau} = \gamma M \wedge B + R(M) \tag{3.17}$$

where γ is the *gyromagnetic ratio*, $B = (B_x, B_y, B_z)$ is the applied magnetic field which decomposes into a *strong polarizing field* $B_z = B_0$ in the z-direction, while B_x, B_y are the components of a *Rf-magnetic field* in the transverse direction and corresponds to the control field and $R(M)$ is the dissipation of the form:

$$\left(-\frac{M_x}{T_2}, -\frac{M_y}{T_2}, -\frac{(M_z - M_0)}{T_1} \right)$$

where T_1, T_2 are the *longitudinal and transverse relaxation parameters* characteristic of the chemical species, e.g. water, fat, blood.

The parameter M_0 is normalized to 1 up to a rescaling of M. We denote $\omega_0 = -\gamma B_0$ the *resonant frequency* and let introduce the control components: $u(\tau) = -\gamma B_y$ and $v(\tau) = -\gamma B_x$. The Bloch equations in the stationary frame can be written in the matrix form:

$$\frac{d}{d\tau} \begin{bmatrix} M_x \\ M_y \\ M_z \end{bmatrix} = \begin{bmatrix} 0 & -\omega_0 & u(\tau) \\ \omega_0 & 0 & -v(\tau) \\ -u(\tau) & v(\tau) & 0 \end{bmatrix} \begin{bmatrix} M_x \\ M_y \\ M_z \end{bmatrix} - \begin{bmatrix} \frac{M_x}{T_2} \\ \frac{M_y}{T_2} \\ \frac{M_z-1}{T_1} \end{bmatrix}. \tag{3.18}$$

The Bloch equations can be represented in a *rotating frame of reference*: $S(\tau) = \exp(\tau \omega \Omega_z)$, $M = S(\tau)q$, $q = (x, y, z)$,

$$\Omega_z = \begin{bmatrix} 0 & -1 & 0 \\ 1 & 0 & 0 \\ 0 & 0 & 0 \end{bmatrix}$$

and introducing the control representation:

$$u_1 = u \cos \omega\tau - v \sin \omega\tau$$
$$u_2 = u \sin \omega\tau + v \cos \omega\tau,$$

one gets the Bloch equations in the moving frame:

$$\frac{d}{d\tau} \begin{bmatrix} x \\ y \\ z \end{bmatrix} = \begin{bmatrix} 0 & -\Delta\omega & u_2 \\ \Delta\omega & 0 & -u_1 \\ -u_2 & u_1 & 0 \end{bmatrix} \begin{bmatrix} x \\ y \\ z \end{bmatrix} - \begin{bmatrix} \frac{x}{T_2} \\ \frac{y}{T_2} \\ \frac{z-1}{T_1} \end{bmatrix} \qquad (3.19)$$

where $\Delta\omega = \omega_0 - \omega$ is the *resonance offset*.

The control is bounded by m, $m = 2\pi \times 32.3$Hz being the experimental intensity of the experiments. Assuming $\Delta\omega = 0$ (resonance), and using the normalized time $t = \tau m$, denoting $\Gamma = 1/mT_2$, $\gamma = 1/mT_1$ and the physical parameters satisfying the constraint: $2\Gamma \geq \gamma \geq 0$, the system is normalized to:

$$\frac{dx}{dt} = -\Gamma x + u_2 z$$
$$\frac{dy}{dt} = -\Gamma y - u_1 z \qquad (3.20)$$
$$\frac{dz}{dt} = \gamma(1 - z) + u_1 y - u_2 x,$$

where $|u| \leq 1$. Moreover since $2\Gamma \geq \gamma \geq 0$, one has that the *Bloch ball* $|q| \leq 1$ is invariant for the dynamics.

This equation describes the evolution of the magnetization vector in NMR. The choice of $\omega = \omega_0$ corresponding to resonance neutralized the existence of the strong polarizing field B_0, except the side effect of a stable unique equilibrium, corresponding to the North pole $N = (0, 0, 1)$ of the Bloch equation for the uncontrolled motion.

In MRI, the situation is more complex due to the spatial position of the spin in the image and one must control an ensemble of spins corresponding to each pixel. Due to this localization they are some *distortions* corresponding to B_0 *and* B_1 *inhomogeneities*. The variation of B_0 producing a resonance offset and $\Delta\omega$ belongs to some intervals, while B_1-inhomogeneity introduces a scaling factor $a_i \geq 0$ depending on the spatial position of the spin in the image modeling the distortion of the amplitude of the control field and the equation transforms into

$$\frac{dx}{dt} = -\Gamma x + a_i u_2 z$$
$$\frac{dy}{dt} = -\Gamma y - a_i u_1 z \qquad (3.21)$$
$$\frac{dz}{dt} = \gamma(1 - z) + a_i(u_1 y - u_2 x).$$

In the general case one must consider both inhomogeneities producing a detuning and amplitude alteration. Note that such distortions cannot be modelized and have to be *experimentally determined*.

To relate Bloch equation to imaging we associate to the amplitude $|q|$ of the normalized magnetization vector a *grey level where $|q| = 1$ corresponds to white while the center of the Bloch ball defined by $|q| = 0$ corresponds to black*.

3.3.2 The Problems

Having introduced the control systems in relation with Bloch equations taking into account B_0 and B_1 inhomogeneities one can present the fundamental problems studied in NMR and MRI.

Saturation Problem

The objective of the saturation problem for a single spin is to bring the magnetization vector q from the North pole $N = (0, 0, 1)$ of the Bloch ball (which is the equilibrium of the free system) to the center $O = (0, 0, 0)$ of the Bloch ball, recalling that $|q|$ corresponds to a grey level where the sphere $|q| = 1$ corresponds to white and $|q| = 0$ to black.

A direct generalization being to consider an *ensemble of spin particles* corresponding to the same chemical species and to bring each spin of this ensemble from the North pole to the center, corresponding to the *multisaturation problem*.

The Contrast Problem

In the contrast problem in NMR called *ideal contrast problem* we consider two pairs of (uncoupled) spin-1/2 systems corresponding to different chemical species, each of them solutions of the Bloch Eq. (3.21) with respective parameters (γ_1, Γ_1) and (γ_2, Γ_2) controlled by the same magnetic field. Denoting each system by $\frac{dq_i}{dt} = f(q_i, \Lambda_i, u)$, $\Lambda_i = (\gamma_i, \Gamma_i)$ and $q_i = (x_i, y_i, z_i)$ is the magnetization vector representing each spin particle, $i = 1, 2$. This leads to the consideration of the system abbreviated as: $\frac{dq}{dt} = f(q, u)$, where $q = (q_1, q_2)$ is the state variable. The constrast problem by saturation is the following optimal control problem: starting from the equilibrium point $q_0 = ((0, 0, 1), (0, 0, 1))$ where both chemical species are white and hence *indistinguishable*, reach in a given transfer time t_f the final state $q_1(t_f)$ corresponding to saturation of the first spin while maximizing $|q_2(t_f)|^2$, the final observed contrast being $|q_2(t_f)|$.

Obvious generalization of the problems in MRI, taking into account B_0 and B_1 inhomogeneities, is to consider in the image an ensemble of N pairs of chemical species, e.g. water or fat, and distributed in the image and the objective is to provide multisaturation of the ensemble of spins of the first species and to reach for the second species a small ball centered at $|q_2(t_f)|$ where $|q_2(t_f)|$ corresponds to the contrast calculated in NMR.

The objective in MRI is to produce a *robust control* taking into account the B_0 and B_1 inhomogeneities.

In the sequel and in order to present the concepts and the theoretical tools, we shall restrict to the saturation problem of a single spin and the contrast problem by saturation in NMR. It is the preliminary step to the analysis of an ensemble of spins which is in the applications treated numerically using adapted software e.g. Bocop and HamPath representative respectively of direct and indirect methods in numeric optimal control.

3.3.3 The Saturation Problem in Minimum Time for a Single Spin

The saturation problem in minimum time was first analyzed in [57] and was an important step to the applications of geometric optimal control to the dynamics of spins particles.

Preliminaries

First of all, since the transfer is from the North pole $N = (0, 0, 1)$ to the center of the Bloch ball $O = (0, 0, 0)$ which belongs to the z-axis of revolution of the system corresponding to polarization the system can be restricted to the two-dimensional plane of the Bloch ball and the control $u = (u_1, u_2)$ reduces to the u_1 component. The system is compactly written as: $\frac{dq}{dt} = F(q) + u_1 G(q)$, while the control is bounded by $|u| \leq 1$ and $q = (y, z)$. We have

$$
F = -\Gamma y \frac{\partial}{\partial y} + \gamma(1 - z) \frac{\partial}{\partial z}
$$
$$
G = -z \frac{\partial}{\partial y} + y \frac{\partial}{\partial z}.
$$
(3.22)

According to the maximum principle an optimal trajectory is a concatenation of bang arcs where $u(t) = \text{sign}\langle p(t), G(q(t))\rangle$ and singular arcs where $\langle p(t), G(q(t))\rangle = 0$. The following Lie brackets are relevant in our analysis. Denoting $\delta = \gamma - \Gamma$, we have

$$
[G, F] = (-\gamma + \delta z) \frac{\partial}{\partial y} + \delta y \frac{\partial}{\partial z}
$$
$$
[[G, F], F] = (\gamma(\gamma - 2\Gamma) - \delta^2 z) \frac{\partial}{\partial y} + \delta^2 y \frac{\partial}{\partial z}
$$
$$
[[G, F], G] = 2\delta y \frac{\partial}{\partial y} + (\gamma - 2\delta z) \frac{\partial}{\partial z}.
$$

Singular Trajectories and Optimality

The singular trajectories are located on the set $S : \det(G, [G, F]) = 0$ which is given by $y(-2\delta z + \gamma) = 0$. Hence it is formed by

- the z-axis of revolutions $y = 0$,
- the horizontal line $z = \gamma/(2\delta)$. This line intersects the Bloch ball $|q| < 1$ when $2\Gamma > 3\gamma$ and moreover z is negative.
  The singular control is given by $D' + u_s D = 0$, where $D = \det(G, [[G, F], G])$ and $D' = \det(G, [[G, F], F])$.
- for $y = 0$, $D = -z(\gamma - 2\delta z)$ and $D' = 0$. The singular control is zero and a singular trajectory is a solution of $\dot{y} = -y$, $\dot{z} = \gamma(1 - z)$ where the equilibrium point $(0, 1)$ is a stable node if $\gamma \neq 0$.
- for $z = \gamma/(2\delta)$, $D = -2\delta y^2$, $D' = y\gamma(2\Gamma - \gamma)$ and $u_s = \gamma(2\Gamma - \gamma)/(2\delta y)$, $2\Gamma - \gamma \geq 0$. Hence along the horizontal direction, the flow: $\dot{y} = -\Gamma y - \gamma^2 \frac{2\Gamma - \gamma}{4\delta^2 y}$ and $|u_s| \to \infty$ when $y \to 0$.

An easy computation gives the following proposition.

Proposition 16 *If $\gamma \neq 0$, the singular control along the singular line is L^1 but not L^2.*

The maximum principle selects the singular line but the high order maximum principle and the so-called generalized Legendre-Clebsch condition [53] has to used to distinguish between small time minimum and maximum solution. It can be easily understood using the two seminal examples:

$$\dot{x} = 1 - u^2, \qquad\qquad \dot{x} = 1 + u^2,$$
$$\dot{y} = u, \; |u| \leq 1, \qquad \dot{y} = u, \; |u| \leq 1$$

where in both case the x-axis is the singular line and is time minimizing in the first case and time maximizing in the second case. The optimality condition takes the following form in our case. Let $D'' = \det(G, F) = \gamma z(z - 1) + \Gamma \gamma^2$. The set $C : D'' = 0$ is the *collinear set*. If $\gamma \neq 0$, this set forms an oval joining the North pole to the center of the Bloch ball and the intersection with the singular line is empty. Denoting $D = \det(G, [[G, F], G])$ the singular lines are fast displacement direction if $DD'' > 0$ and slow if $DD'' < 0$. From this condition, one deduces that the z-axis of revolution is fast if $1 > z > z = \gamma/(2\delta)$ and slow if $z = \gamma/(2\delta) > z > -1$, while the horizontal singular line is fast.

From the analysis we deduce first

Lemma 7 *If the condition $2\Gamma > 3\gamma$ is not satisfied the horizontal singular line doesn't intersect the Bloch ball $|q| < 1$ and the optimal solution is the standard inversion sequence used in practices: apply a bang control to steer $(0, 1)$ to $(0, -*)$. Followed by $u = 0$ to relax the system to $(0, 0)$ along the z-axis.*

If $2\Gamma > 3\gamma$ the existence of the fast displacement horizontal line will determine the optimal policy. First of all observe that since $u_S \to \infty$, when $q \to 0$ along this line, it is saturating the constraint $|u| < 1$ at a point of this line. Hence this line has to be quitted before this point. The exact exit point is determined by the maximum principle because such point has to be a switching point at both extremities for the corresponding bang arc. Such an arc is called *a bridge* (Fig. 3.1).

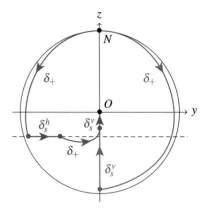

Fig. 3.1 *(left)* Time minimal solution compared with *(right)* inversion sequence

Note that in this analysis we assume that the applied RF-field is large enough, which correspond to the experimental situation.

We deduce the following theorem, see [57] for further details.

Theorem 12 *If* $2\Gamma > 3\gamma$, *in the time minimal saturation problem is of the form:* $\delta_+\delta_s^h\delta_+\delta_s^v$, *concatenating the bang arc to quit the North pole to the horizontal singular line, followed by the bridge and relaxation to* 0 *along the z-axis of revolution.*

Remark 3.1 The bridge can be empty and in this case the optimal policy is $\delta_+\delta_s^v$.

This gives a complete solution to the saturation problem using a careful geometric analysis to understand the interaction between the two singular lines. Moreover a similar analysis leads to a complete understanding of the time minimum synthesis to transfer any point of the Bloch ball to the center.

Extension of this type of results to an ensemble of two or more spins is an important issue. The complexity is related to the analysis of singular extremals at two levels. First of all, in general the symmetry of revolution due to z-polarization cannot be invoked to reduced the bi-inputs case to the single single-input case. Secondly, in dimension ≥ 3, the analysis of the singular flow even in the single-input case is a complicated task. Next, we shall present this complexity in the contrast problem and present some achievements.

3.3.4 The Maximum Principle in the Contrast Problem by Saturation

The system is written as:

$$\dot{q} = F_0(q) + u_1 F_1(q) + u_2 F_2(q), \quad |u| \leq 1$$

where $q = (q_1, q_2) \in \{ |q_1| \leq 1, \ |q_2| \leq 1 \}$ and q_1, q_2 represents the normalized magnetization vector of the first and second spin, $q_i = (x_i, y_i, z_i)$, $i = 1, 2$. Using the notation of the Sect. 3.2 for a Mayer problem, the cost function is $c(q(t_f)) = -|q_2(t_f)|^2$ and the final boundary condition is $F(q(t_f)) = q_1(t_f) = 0$. Splitting the adjoint vector into $p = (p_1, p_2) \in \mathbb{R}^3 \times \mathbb{R}^3$, the transversality condition is:

$$p_2(t_f) = -2p^0 q_2(t_f), \ p^0 \leq 0$$

and if $p^0 \neq 0$ it can be normalized to $p^0 = -1/2$.

We denote $z = (q, p)$, $H_i = \langle p, F_i(q) \rangle$, $i = 0, 1, 2$, the Hamiltonian lift of the system $\dot{z} = \overrightarrow{H}_0 + \sum_{i=1}^2 \overrightarrow{H}_i(z)$. If $(H_1, H_2) \neq 0$, the maximization condition of the maximum principle leads to the following parametrization of the controls

$$u_1 = \frac{H_1}{\sqrt{H_1^2 + H^2}}, \qquad u_2 = \frac{H_2}{\sqrt{H_1^2 + H_2^2}}.$$

Define the *switching surface*:

$$\Sigma \ : \ H_1 = H_2 = 0.$$

Plugging such u into the pseudo-Hamiltonian gives the true Hamiltonian: $H_n = H_0 + \sqrt{H_1^2 + H_2^2}$. The corresponding extremal solutions are called zero.

Besides those generic extremals, additional extremals are related to Lie algebraic properties of the system and a careful analysis is the key factor to determine the properties of the optimal solutions.

Lie Bracket Computations

Due to the bilinear structure of the Bloch equations, Lie brackets can be easily computed, which is crucial in our analysis.

Recall that the Lie bracket of two vectors fields F, G is computed with the convention

$$[F, G](q) = \frac{\partial F}{\partial q}(q)G(q) - \frac{\partial G}{\partial q}(q)F(q)$$

and if H_F, H_G are the Hamiltonian lifts, recall that the Poisson bracket is

$$\{H_F, H_G\}(z) = dH_F(\overrightarrow{G})(z) = H_{[F,G]}(z).$$

To simplify the computation, each spin system is lifted on the *semi-direct Lie product* $GL(3, \mathbb{R}) \times \mathbb{R}^3$ acting on the q-space using the action $(A, a).q = Aq + a$. The Lie bracket computation rule is $((A, a), B, b) = ([A, B], Ab - Ba)$ where $[A, B] = AB - BA$.

Introducing $F_0 = (A_0, a_0)$, with $A_0 = \text{diag}(-\Gamma_1, -\Gamma_1, -\gamma_1, -\Gamma_2, -\Gamma_2, -\gamma_2)$ and $a_0 = (0, 0, \gamma_1, 0, 0, \gamma_2)$ whereas the control fields (F_1, F_2) are identified to

$B_1 = \mathrm{diag}(C_1, C_1)$ and $B_2 = \mathrm{diag}(C_2, C_2)$ where C_1, C_2 are the antisymmetric matrices $C_1 = E_{32} - E_{23}, C_2 = (E_{13} - E_{31})$ with $E_{ij} = (\delta_{ij})$ (Kronecker symbol). See [24] for more details.

Next, we present in details the Lie brackets needed in our computations, each entry form by a couple (v_1, v_2) and we use the notation omitting the indices. We set $\delta = \gamma - \Gamma$.

- *Length 1*:

$$F_0 = (-\Gamma x, -\Gamma y, \gamma(1 - z))$$
$$F_1 = (0, -z, y)$$
$$F_2 = (z, 0, -x).$$

- *Length 2*:

$$[F_0, F_1] = (0, \gamma - \delta z, -\delta y)$$
$$[F_0, F_2] = (-\gamma + \delta z, 0, \delta x)$$
$$[F_1, F_2] = (-y, x, 0).$$

- *Length 3*:

$$[[F_1, F_2], F_0] = 0$$
$$[[F_1, F_2], F_1] = F_2$$
$$[[F_1, F_2], F_2] = -F_1$$
$$[[F_0, F_1], F_1] = (0, -2\delta y, -\gamma + 2\delta z)$$
$$[[F_0, F_1], F_2] = (\delta y, \delta x, 0) = [[F_0, F_2], F_1]$$
$$[[F_0, F_2], F_2] = (-2\delta x, 0, 2\delta z - \gamma)$$
$$[[F_0, F_1], F_0] = (0, -\gamma(\gamma - 2\Gamma) + \delta^2 z, -\delta^2 y)$$
$$[[F_0, F_2], F_0] = (\gamma(\gamma - 2\Gamma) - \delta^2 z, 0, \delta^2 x).$$

3.3.5 Stratification of the Surface $\Sigma : H_1 = H_2 = 0$ and Partial Classification of the Extremal Flow Near Σ

Let $z = (q, p)$ be a curve solution of $\overrightarrow{H}_0 + u_1\overrightarrow{H}_1 + u_2\overrightarrow{H}_2$. Differentiating H_1 and H_2 along such a solution, one gets:

$$\begin{aligned}\dot{H}_1 &= \{H_1, H_0\} + u_2\{H_1, H_2\}\\ \dot{H}_2 &= \{H_2, H_0\} + u_1\{H_2, H_1\}.\end{aligned} \tag{3.23}$$

Hence we have:

Proposition 17 Let $z_0 \subset \Sigma_1 = \Sigma \setminus \{H_1, H_2\} = 0$ and define the control u_s by:

$$u_s(z) = \frac{(-\{H_0, H_2\}(z), \{H_0, H_1\}(z))}{\{H_1, H_2\}(z)}, \tag{3.24}$$

and plugging such u_s into H defines the true Hamiltonian

$$H_s(z) = H_0(z) + u_{s,1}(z)H_1(z) + u_{s,2}(z)H_2(z)$$

which parameterizes the singular solutions of the bi-input system contained in Σ_1.

This gives the first stratum of the surface Σ. Moreover, the behaviors of the extremals of order zero near a point z_0 of Σ_1 can be easily analyzed using (3.23) and a nilpotent model where all Lie brackets at $z_0 \in \Sigma_1$ of length ≥ 3 are zero. Denoting:

$$\{H_1, H_0\}(z_0) = a_1, \quad \{H_2, H_0\}(z_0) = a_2, \quad \{H_2, H_1\}(z_0) = b$$

and using polar coordinates $H_1 = r \cos \theta$, $H_2 = r \sin \theta$, then (3.23) becomes:

$$\begin{aligned} \dot{r} &= a_1 \cos \theta + a_2 \sin \theta \\ \dot{\theta} &= \frac{1}{r}(b - a_1 \sin \theta + a_2 \cos \theta). \end{aligned} \tag{3.25}$$

To analyze this equation, we write:

$$a_1 \sin \theta - a_2 \cos \theta = A \sin(\theta + \phi)$$

with $A \tan \phi = -a_2/a_1$, $A = \sqrt{a_1^2 + a_2^2}$. Hence the equation $\dot{\theta} = 0$ leads to the relation

$$A \sin(\theta + \phi) = b,$$

which has two distinct solutions on $[0, 2\pi[$ denoted θ_0, θ_1 if and only if $A > |b|$, one solution if $A = |b|$ and zero solution if $|A| < |b|$. Moreover $\theta_1 - \theta_0 = \pi$ if and only if $b = 0$. Plugging θ_0, θ_1 in (3.25), one gets two solutions of (3.25). Hence we deduce:

Lemma 8 If $\sqrt{a_1^2 + a_2^2} > |b|$ and $b \neq 0$, we have a broken extremal formed by concatenating two extremals of order zero at each point z_0 of Σ_1.

At such a point z_0 of Σ_1, the singular control (3.24) is such that

$$u_{s,1}^2 + u_{s,2}^2 = \frac{a_1^2 + a_2^2}{b^2} > 1$$

and hence is not admissible.

Next we analyze more degenerated situations and one needs the following concept.

Goh Condition

Higher order necessary optimality conditions along singular extremals in the bi-input case are related to finiteness of the index of the quadratic forms associated with the second order derivative [22] known as *Goh condition* which is the relation:

$$\{H_1, H_2\} = 0. \tag{3.26}$$

Using $H_1 = H_2 = \{H_1, H_2\} = 0$ and (3.23), one gets the additional relations:

$$\{H_1, H_2\} = \{H_0, H_1\} = \{H_0, H_2\} = 0. \tag{3.27}$$

Then differentiating again along a solution leads to the relations:

$$\{\{H_1, H_2\}, H_0\} + u_1\{\{H_1, H_2\}, H_1\} + u_2\{\{H_1, H_2\}, H_2\} = 0 \tag{3.28}$$

$$\begin{cases} \{\{H_0, H_1\}, H_0\} + u_1\{\{H_0, H_1\}, H_1\} + u_2\{\{H_0, H_1\}, H_2\} = 0 \\ \{\{H_0, H_2\}, H_0\} + u_1\{\{H_0, H_2\}, H_1\} + u_2\{\{H_0, H_2\}, H_2\} = 0 \end{cases} \tag{3.29}$$

This leads in general to *three* relations to compute *two* control components, and for a generic system such conditions are not satisfied [35], but in our case, according to Lie brackets computations, we have:

Lemma 9 *If* $H_1 = H_2 = 0$, *one has*

$$\{\{H_1, H_2\}, H_0\} = \{\{H_1, H_2\}, H_1\} = \{\{H_1, H_2\}, H_2\} = 0$$

and (3.28) *is satisfied.*

The Eq. (3.29) are then written: $\tilde{A} + \tilde{B}u$ and if $\det(\tilde{B}) \neq 0$, the corresponding singular control is given by:

$$u'_s(z) = -\tilde{B}^{-1}(z)\tilde{A}(z) \tag{3.30}$$

Using the relations:

$$H_1 = H_2 = \{H_1, H_2\} = \{H_0, H_1\} = \{H_0, H_2\} = 0,$$

the vector p is orthogonal to $F_1, F_2, [F_1, F_2], [F_0, F_1], [F_0, F_2]$. Introducing:

$$A = \begin{pmatrix} A_1 \\ A_2 \end{pmatrix}, \quad B = \begin{pmatrix} B_1 & B_3 \\ B_2 & B_4 \end{pmatrix}, \quad C = (F_1, F_2, [F_1, F_2], [F_0, F_1], [F_0, F_2]),$$

with

$$A_1 = \det(C, [[F_0, F_1], F_0]), \quad A_2 = \det(C, [[F_0, F_2], F_0]),$$

and

$$B_1 = \det(C, [[F_0, F_1], F_1]), \quad B_2 = \det(C, [[F_0, F_2], F_1]),$$
$$B_3 = \det(C, [[F_0, F_1], F_2]), \quad B_4 = \det(C, [[F_0, F_2], F_2]),$$

the relation (3.29) leads to:

$$A + Bu = 0,$$

and if $\det B \neq 0$, one gets the singular control given by the feedback:

$$u_s'(q) = -B^{-1}(q)A(q) \tag{3.31}$$

and the associated vector field:

$$Q_s' = F_0 + u_{s,1}' F_1 + u_{s,2}' F_2.$$

Moreover, the singular control has to be admissible: $|u_s'| \leq 1$. We introduce the stratum:

$$\Sigma_2 : H_1 = H_2 = \{H_1, H_2\} = \{H_0, H_1\} = \{H_0, H_2\} \setminus \det \tilde{B} = 0.$$

Hence we have:

Lemma 10   *1. On the stratum Σ_2, there exists singular extremals satisfying Goh condition where the singular control is given by the feedback (3.30).*
  2. For the contrast problem:

$$\det B = (x_1 y_2 - x_2 y_1)^4 (\delta_1 - \delta_2)(2\delta_1 z_1 - \gamma_1)(2\delta_2 z_2 - \gamma_2)$$
$$\left(2(\delta_1^2 \gamma_2 z_1 - \delta_2^2 \gamma_1 z_2) - \gamma_1 \gamma_2 (\delta_1 - \delta_2) - 2\delta_1 \delta_2 (\gamma_1 z_2 - \gamma_2 z_1)\right), \tag{3.32}$$

The behaviors of the extremals of order zero near a point $z_0 \in \Sigma_2$ is a complicated problem. Additional singular extremals can be contained in the surface:

$$\Sigma_3 : H_1 = H_2 = \{H_1, H_2\} = \{H_0, H_1\} = \{H_0, H_2\} = \det \tilde{B} = 0,$$

and they can be computed using the property that the corresponding control has to force the surface $\det B = 0$ to be invariant. *Some have an important meaning, due to the symmetry of revolution of the Bloch equations.* They correspond to control the system, imposing $u_2 = 0$. In this case, one can restrict the system to

$$\mathbf{Q} = \{q = (q_1, q_2) \in \mathbb{R}^n : |q_1| \leq 1, |q_2| \leq 1, x_1 = x_2 = 0\}.$$

The computations of the corresponding extremals amount to replace in the relations; H_2 by εH_2 and to impose $\varepsilon = 0$. The remaining relations are then:

$$H_1 = \{H_0, H_1\} = 0$$

and from (3.29) one gets the relations:

$$\{\{H_0, H_1\}, H_0\} + u_{1,s}\{\{H_0, H_1\}, H_1\} = 0, \tag{3.33}$$

and thus, this defines the singular control:

$$u_{1,s} = -\frac{\{\{H_0, H_1\}, H_0\}}{\{\{H_0, H_1\}, H_1\}} \tag{3.34}$$

and the associated Hamiltonian $H_{1,s} = H_0 + u_{1,s}H_1$. We have the following result:

Proposition 18 *The singular extremals of the single-input case with $u_2 \equiv 0$ are extremals of the bi-input case with the additional condition: $x_1 = p_{x_1} = x_2 = p_{x_2} = 0$.*

Moreover from the geometric interpretation of the maximum principle for a Mayer problem, in order to be optimal the generalized Legendre-Clebsch condition has to be satisfied:

$$\frac{\partial}{\partial u_1}\frac{d^2}{dt^2}\frac{\partial H}{\partial u_1} = \{H_1, \{H_1, H_0\}\}(z) \leq 0. \tag{3.35}$$

Observe that if we impose $u_2 = 0$, the classification of the extremals near the switching surface, which reduces to $H_1 = 0$, is a standard problem [54].

Finally, another important property of the extremal flow, again a consequence of the symmetry of revolution is given next, in relation with Goh condition. It is a consequence of Noether integrability theorem.

Proposition 19 *In the contrast problem, for the Hamiltonian vector field $\overrightarrow{H}_n$ whose solutions are extremals of order zero, the Hamiltonian lift $H(z) = \{H_1, H_2\}(z) = (p_{x_1}y_1 - p_{y_1}x_1) + (p_{x_2}y_2 - p_{y_2}x_2)$ is a first integral.*

Exercise 3.2 (*Generalization to the case of B_1 and B_0 inhomogeneities*) It is interesting to compare to the case of an ensemble of two spins of the same spin particle with B_0 and B_1 inhomogeneities which is left to the reader. More precisely:

- B_1-*inhomogeneities*
  In this case, the control directions of the second spin are relaxed by a factor and the Lie brackets computations can be used to stratified. It can be applied to the multisaturation problem.

- B_0-inhomogeneities

  In this case the vector field F_0 of the second spin contains a non-zero detuning. Clearly this introduces modifications in the Lie brackets computations. Again it can be applied to multisaturation problem. It explains the following phenomenon: *in the precense of detuning both controls (u_1, u_2) have to be used.*

Next, motivating by the fact that due to the symmetry of revolution and the observed numerical experiments, we shall restrict our study to the single-input case. It is an important theoretical step since we can reduce the analysis of the singular flow for a 4-dimensional system with one input vs a 6-dimensional system. This complexity will be illustrated by the computations presented next.

3.3.6   The Classification of the Singular Extremals and the Action of the Feedback Group

Preliminairies

Restricting to the single input case, the research program concerning the contrast problem or the multisaturation problem for an ensemble of two spins is clear.

Saturation Problem for a Single Spin and Bridge Phenomenon

In the case of a single spin the complete geometric analysis requires the computations of the two singular line and the understanding of the singularity associated with their intersection, which causes the saturation of the singular control and the occurrence of a bang arc called a bridge to connect both singular arcs. This phenomenon generalizes to higher dimension and it tells you that the analysis of the singular flow codes all the information of the optimal solution which is a sequence of arcs of the form $\delta_\pm \delta_S \delta_\pm \cdots \delta_S$, where $\delta_\pm$ denotes bang arcs with $u = \pm 1$, while δ_S are singular arcs.

This will be presented in details next, in relation with the action of the feedback group.

Computations of the Singular Flow

Consider a control system of the form:

$$\frac{dq}{dt} = F(q) + uG(q), \quad q \in \mathbb{R}^n$$

and relaxing the control constraints: $u \in \mathbb{R}$. Denoting H_F and H_G the Hamiltonian lifts of F and G, if the denominator is not vanishing, a singular control is given by:

$$u_S(z) = -\frac{\{\{H_G, H_F\}, H_F\}(z)}{\{\{H_G, H_F\}, H_G\}(z)}. \tag{3.36}$$

Plugging such u_S into the pseudo-Hamiltonian one gets the true Hamiltonian: $H_S = H_F + u_S H_G$ and the singular extremals are solutions of the *constrained Hamiltonian equation*:

$$\frac{dz}{dt} = \overrightarrow{H}_S(z), \; z \in \Sigma' \; : \; H_G = \{H_G, H_F\} = 0.$$

This set of equations defines a Hamiltonian vector field on the surface

$$\Sigma' \setminus \{\{H_G, H_F\}, H_G\} = 0$$

, restricting the standard symplectic from $\omega = dp \wedge dq$.

We use the notation $\mathscr{D} = \{\{H_G, H_F\}, H_G\}$ and $\mathscr{D}' = \{\{H_G, H_F\}, H_F\}$. The differential Eq. (3.36) can be desingularized using the time reparametrization

$$ds = dt / \mathscr{D}(z(t))$$

which amounts to analyze the one dimensional foliation.

We get the system:

$$\frac{dq}{ds} - \mathscr{D}F - \mathscr{D}'F, \; \frac{dp}{ds} = -p\left(\mathscr{D}\frac{\partial F}{\partial q} - \mathscr{D}'\frac{\partial G}{\partial q}\right)$$

restricted to the surface Σ'.

In the contrast problem, since the state space is of dimension four, using the two constraints $H_G = \{H_G, H_F\} = 0$ and the homogeneity with respect to p, Eq. (3.36) can be reduced to the explicit form:

$$\frac{dq}{dt} = F(q) - \frac{\mathscr{D}'(q, \lambda)}{\mathscr{D}(q, \lambda)} G(q)$$

where λ is a one-dimension time-dependant parameter whose dynamics is deduced from the adjoint equation.

Using the previous remark, the optimal problem can be analyzed by understanding the behavior of the corresponding trajectories and the singularities of the flow near the set $\mathscr{D} = 0$, which codes the switching sequence.

This is a very complicated task, in particular because the system is depending upon four parameters and simplifications have to be introduced to simplify this task. Two simplifications can be introduced. First, we can restrict to some specific parameters corresponding to some experimental cases. For instance, in the water case, saturation of a single spin amounts to the standard inversion sequence. Second, a projection of the singular flow which is physically relevant can be introduced. A natural choice is to consider the case where the transfer time t is not fixed. Then according to the maximum principle this leads to the additional constraint: $M = \underset{u(.)}{\text{Max}}\, H_F + u H_G = 0$, which gives in the singular case the additional constraint:

$H_F(z) = 0$. This case is called the *exceptional case* using the terminology of [29].

With this constraint, the adjoint vector can be eliminated and the singular control in this exceptional case is the feedback:

$$u_S^e = -\frac{D'(q)}{D(q)},$$

where $D = \det(F, G, [G, F], [[G, F], G])$, $D' = \det(F, G, [G, F], [[G, F], F])$ with the corresponding vector field X^e defined by

$$\frac{dq}{dt} = F(q) - \frac{D'(q)}{D(q)} G(q)$$

which can again be desingularized using the reparametrization $ds = dt/D(q(t))$ and this gives the smooth vector field

$$X_r^e = DF - D'G.$$

Feedback Classification

Definition 41 Let E and F be two $\mathbb{R}$-vector spaces and let $\mathfrak{G}$ be a group acting linearly on E and F. A homomorphism $\mathfrak{X} : \mathfrak{G} \to \mathbb{R} \setminus \{0\}$ is called a *character*. A *semi-invariant of weight* $\mathfrak{X}$ is a map $\lambda : E \to \mathbb{R}$ such that for all $g \in \mathfrak{G}$ and all $x \in E, \lambda(g, x) = \mathfrak{X}(g)\lambda(x)$; it is an *invariant* if $\mathfrak{X} = 1$. A map $\lambda : E \to F$ is a *semi-covariant of weight* $\mathfrak{X}$ if for all $g \in \mathfrak{G}$ and for all $x \in E, \lambda(g.x) = \mathfrak{X}(g)g.\lambda(x)$; it is called a *covariant* if $\mathfrak{X} = 1$.

More about invariant theory can be found in [39].

The key concept in analyzing the role of relaxation parameters in the control problem is the action of the *feedback group* $\mathfrak{G}$ on the set of systems. We shall restrict our presentation to the single-input case and we denote $\mathscr{C} = \{F, G\}$ the set of such (smooth) systems on the state space $Q \simeq \mathbb{R}^n$, see [20] for the details.

Definition 42 Let $(F, G), (F', G')$ be two elements of $\mathscr{C}$. They are called feedback equivalent if there exist a smooth diffeomorphism φ of $\mathbb{R}^n$ and a feedback $u = \alpha(q) + \beta(q)v$, β invertible such that:

• $F' = \varphi * F + \varphi * (G\alpha)$, $G' = \varphi * (G\beta)$.

where $\varphi * z$ denotes the image of the vector field.

Definition 43 Let $(F, G) \in \mathscr{C}$ and let λ_1 be the map which associated the constrained Hamiltonian vector field $(\overrightarrow{H}_S, \Sigma')$ (see (3.36)) to (F, G). We define the action of (φ, α, β) of $\mathfrak{G}$ on $(\overrightarrow{H}_S, \Sigma')$ to be the action of the symplectic change of coordinates:

$$\overrightarrow{\varphi} : q = \varphi(Q), \quad p = P\frac{\partial\varphi}{\partial x}^{-1}$$

in particular the feedback acts trivially.

Theorem 13 ([20]) *The mapping λ_1 is a covariant.*

Next, we detail the induced action restricting to exceptional singular trajectories when dim $Q = 4$.

The Exceptional Singular Trajectories and the Feedback Classification
Notation. Let φ be a diffeomorphism of Q. Then φ acts on the mapping $F : Q \to R$ according to $\varphi.F = F \circ \varphi$ and on vector fields X as $\varphi.X = \varphi * X$ (image of X): this corresponds to the action on tensors.

The feedback group acts on the vector field X^e by change of coordinates only and this can be checked as a consequence of the following lemma.

Lemma 11 • $D^{F+\alpha G, \beta G} = \beta^4 D^{F,G}$.
• $D'^{F+\alpha G, \beta G} = \beta^3 \left(D'^{F,G} + \alpha D^{F,G} \right)$.
• $D^{\varphi * F, \varphi * G}(q) = \det \left(\frac{\partial \varphi}{\partial q}^{-1} \right) D^{F,G}(\varphi(q))$.
• $D'^{\varphi * F, \varphi * G}(q) = \det \left(\frac{\partial \varphi}{\partial q}^{-1} \right) D'^{F,G}(\varphi(q))$.

From which we deduce the following crucial result in our analysis.

Theorem 14 *We have the following:*

• $\lambda_2 : (F, G) \to X^e$ *is a covariant.*
• $\lambda_3 : (F, G) \to D$ *is a semi-covariant.*
• $\lambda_4 : (F, G) \to X_r^e = DF - D'G$ *is a semi-covariant.*

The classification program. Having introduced the concepts and results, the contrast problem is related to the following classification program (up to change of coordinates)

• Classification of the vector fields $X_r^e = DF - DF'$ and the surfaces: $D = 0, D = D' = 0$.

Interpretation.

• The singular control is $u_S^e = -\frac{D'}{D}$ and will explode at $D = 0$ except if $D' = 0$, taking into account the (non isolated) singularities of X_r^e (if $D = D' = 0$, $X_r^e = 0$).

Collinear Set. The collinear set of F, G is a feedback invariant which has also an important meaning in our classification.

Remark 3.2 In our classification program we use semi-covariants and in the set of parameters $\Lambda = (\gamma_1, \Gamma_1, \gamma_2, \Gamma_2)$ it amounts to work in the projective space. It is also clear from our reparametrization of time.

Now, the problem is to test the computational limits of our program which is clearly:

- Compute the surfaces $D = 0$, $D = D' = 0$,
- Compute the equilibrium points of X_r^e.

Clearly, in the framework of computational methods in real algebraic geometry it is a complicated task which has been achieved in two cases.

- The multisaturation problem of two spins taking into account B_1-inhomogeneity.
- The contrast problem when the first spin system corresponds to water ($\gamma_1 = \Gamma_1$). The second problem has application in in vivo, where the parameters are varying, in particular in the brain.

We shall present the results in details in the first case.

3.3.7   Algebraic Classification in the Multisaturation of Two Spins with B_1-inhomogeneity

The point $N = ((0, 1), (0, 1))$ is a singular point of X_e^r and under a translation N is taken as the origin of the coordinates. We have:

$$F_0 = (-\Gamma y_1, -\gamma z_1, -\Gamma y_2, -\gamma z_2),$$
$$F_1 = ((-(z_1 + 1), y_1), (1 - \varepsilon)(-(z_2 + 1), y_2))$$

where $(1 - \varepsilon)$ denotes the control rescaling of the second spin.

We have $D = (1 - \varepsilon)\tilde{D}$, where $\tilde{D}$ is a quadric which decomposes into $h_2 + h_3 + h_4$ where h_i are the homogeneous part of degree i:

$h_2 = (2\Gamma - \gamma)\bar{h}_2$

$\bar{h}_2 = \Gamma (2\Gamma - \gamma) ((\varepsilon - 1) y_1 + y_2)^2 + \gamma^2 (\varepsilon - 1)^2 z_1^2 - \gamma^2 (2 - 2\varepsilon + \varepsilon^2) z_2 z_1 + \gamma^2 z_2^2$

$h_3 = 2(\gamma - \Gamma)\bar{h}_3$

$\bar{h}_3 = (\gamma - 2\Gamma)(\gamma + 2\Gamma (\varepsilon - 1)^2) z_2 y_1^2 + (\gamma - 2\Gamma)(\gamma + 2\Gamma)(\varepsilon - 1)(y_2 z_1 + z_2 y_2) y_1 -$
$\quad \gamma^2 \varepsilon (\varepsilon - 2) z_2 z_1^2 + ((\gamma - 2\Gamma)(2\Gamma + (\varepsilon - 1)^2 \gamma) y_2^2 + \gamma^2 \varepsilon (\varepsilon - 2) z_2^2) z_1$

$h_4 = 4(\gamma - \Gamma)^2 \bar{h}_4$

$\bar{h}_4 = (\gamma + (\varepsilon - 1)^2 \Gamma) z_2^2 y_1^2 + 2 (\varepsilon - 1)(\gamma + \Gamma) z_2 y_2 z_1 y_1 + (\Gamma + (\varepsilon - 1)^2 \gamma) y_2^2 z_1^2$

$D' = 2\gamma^2 (\Gamma - \gamma)(2\Gamma - \gamma)(1 - \varepsilon)(z_1 - z_2)((\varepsilon - 1) z_1 y_2 + z_2 y_1).$

In particular we deduce (compare with [23] in the contrast problem):

Proposition 20 *The quadric D' reduces to a cubic form which is factorized into a linear and a quadratic (homogeneous) forms.*

Singular Analysis

We assume $\gamma > 0$ and $2\Gamma > 3\gamma$. It implies $\gamma \neq \Gamma$ and $\gamma \neq 2\Gamma$. The main result is the following:

Theorem 15 *Provided $\varepsilon \neq 1$ the equilibrium points of $X_e^r = DF_0 - D'F_1$ are all contained in $\{D = D' = 0\}$.*

A simple proof exists, but we present a method based on symbolic computation and Gröbner basis.

Proof Obviously, every point of $\{D = 0\} \cap \{D' = 0\}$ is a singularity of X_e^r.

Conversely, let us assume $\varepsilon \neq 1$. We first divide X_e^r by $1 - \varepsilon$. We still assume that $\Gamma \neq 0$. We consider the equations $\{(X_e^r)_{y_1} = 0, (X_e^r)_{z_1} = 0, (X_e^r)_{y_2} = 0, (X_e^r)_{z_2} = 0\}$ and remark that the last third are dividable by γ. By homogeneity, changing γ into $\gamma\Gamma$, we get rid of Γ. So we may assume $\Gamma = 1$. The resulting system is denoted Σ_r. We add the two polynomials $((\varepsilon - 1)\, z_1\, y_2 + z_2\, y_1)\, a_1 - 1$ and $(z_1 - z_2)\, a_2 - 1$, and the polynomials $\gamma g - 1$, $(\gamma - 1)g_1 - 1$, $(\gamma - 2)g_2 - 1$. We denote $\tilde{\Sigma}_r$ this new system, involving four new variables g_1, g_2, a_1, a_2. We compute a Gröbner basis with total degree with reverse lexicographic order on $(y_1, y_2, z_1, z_2, \varepsilon, g, g_1, g_2, a_1, a_2)$ and get $\{1\}$. Hence, provided γ is different from $0, 1, 2$, there is no singular point of X_e^r outside of $\{D = 0\} \cap \{D' = 0\}$.

The remaining of the section is devoted to the singularity resolution. From the factorized form of D' (Proposition 20) we get:

Proposition 21 $\{D = 0\} \cap \{D' = 0\}$ *is an algebraic variety of algebraic dimension 2 whose components are located in the hyperplane $z_1 = z_2$ and in the hypersurface $(\varepsilon - 1)\, z_1 y_2 + z_2 y_1 = 0$.*

These components are studied in the following analysis, and explicitly expressed in Lemmas 12, 13, 14, 15.

- Case A: components of $\{D = 0\} \cap \{D' = 0\}$ in $z_1 = z_2$.
Under the constraint $z_1 = z_2$, we have a factorization $\tilde{D} = p_1\, p_2$ with:

$$p_1 = 2\,(\gamma - \Gamma)\, z_1 + \gamma - 2\,\Gamma$$

and:

$$
\begin{aligned}
p_2 = &\left(2\,(\gamma - \Gamma)\left(\gamma + (\varepsilon - 1)^2\,\Gamma\right) z_1 + \Gamma\,(\varepsilon - 1)^2\,(\gamma - 2\,\Gamma)\right) y_1^2 + \\
&\left(4\,(\gamma - \Gamma)(\gamma + \Gamma)(\varepsilon - 1)\, z_1 + 2\,\Gamma\,(\varepsilon - 1)(\gamma - 2\,\Gamma)\right) y_2\, y_1 + \\
&\left(2\,(\gamma - \Gamma)\left(\Gamma + (\varepsilon - 1)^2\,\gamma\right) z_1 + \Gamma\,(\gamma - 2\,\Gamma)\right) y_2^2.
\end{aligned}
$$

The first polynomial has one root $z_1 = z_{\gamma,\Gamma}$

$$z_{\gamma,\Gamma} = \frac{1}{2}\,\frac{2\,\Gamma - \gamma}{\gamma - \Gamma}$$

which corresponds to the plane-solution $\{(y_1, z_{\gamma,\Gamma}, y_2, z_{\gamma,\Gamma}), \ (y_1, y_2) \in \mathbb{R}^2\}$.
We put:

$$d_2(y_1, y_2) = \left(\gamma + (\varepsilon - 1)^2 \, \Gamma\right) y_1^2 + 2 \, (\varepsilon - 1) \, (\gamma + \Gamma) \, y_2 \, y_1 + \left(\Gamma + (\varepsilon - 1)^2 \, \gamma\right) y_2^2.$$

The discriminant of d_2 with respect to y_1 is $-4 \, (\varepsilon - 2)^2 \, \gamma \, \Gamma \, \varepsilon^2 \, y_2^2$ which is strictly
negative provided $\varepsilon \neq 0$. So d_2 is non-zero outside $y_1 = y_2 = 0$.
So, provided $y_1^2 + y_2^2 \neq 0$, $d_2 \neq 0$, and $p_2 = 0$ is solved with respect to z_1. We
get $z_1 = r_2(y_1, y_2)$ with

$$r_2(y_1, y_2) = \frac{\Gamma \, (2\,\Gamma - \gamma) \, ((\varepsilon - 1) \, y_1 + y_2)^2}{2 \, (\gamma - \Gamma) \, d_2(y_1, y_2)}$$

and $(y_1, r_2(y_1, y_2), y_2, r_2(y_1, y_2))$ (defined for $(y_1, y_2) \neq (0, 0)$) vanishes both D
and D'.
Finally, if $y_1 = y_2 = 0$, we have the solution $(0, z, 0, z), \quad z \in \mathbb{R}$.
We summarize the case $z_1 = z_2$ in:

Lemma 12 $\{D = 0\} \cap \{D' = 0\} \cap \{z_1 = z_2\}$ *is the union of an affine plane* $z_1 = z_2 = z_{\gamma,\Gamma}$, *a rational surface* $z_1 = z_2 = r_2(y_1, y_2)$ *(defined for* $(y_1, y_2) \neq (0, 0)$*), and the line* $\{(0, z, 0, z), \ z \in \mathbb{R}\}$.

- Case B: components of $\{D = 0\} \cap \{D' = 0\}$ in $(\varepsilon - 1) \, z_1 y_2 + z_2 y_1 = 0$.

  - Assume first that $y_1 = 0$ and $z_1 \neq z_2$ We have $z_1 y_2 = 0$.
    · If $y_1 = z_1 = 0$, then:

$$\tilde{D} = (\gamma - 2\,\Gamma) \left(\Gamma \, (2\,\Gamma - \gamma) \, y_2^2 + \gamma^2 z_2^2\right)$$

   Since $2\Gamma > \gamma$, $\{\tilde{D} = 0\} \cap \{y_1 = z_1 = 0\}$ corresponds to the North pole N.
    · If $y_1 = y_2 = 0$, then let us put

$$d_1(z_1) = 2\,\varepsilon \, (\varepsilon - 2) \, (\gamma - \Gamma) \, z_1 + 2\,\Gamma - \gamma.$$

We have:

$$\tilde{D} = \gamma^2(z_2 - z - 1)(d_1(z_1)z_2 - (\varepsilon - 1)^2 \, (2\,\Gamma - \gamma) \, z_1.$$

Observe that the polynomial d_1 vanishes if and only if z_1 equals $\tilde{z}_{\gamma,\Gamma}$ with

$$\tilde{z}_{\gamma,\Gamma} = \frac{1}{2} \frac{\gamma - 2\,\Gamma}{\varepsilon \, (\varepsilon - 2) \, (\gamma - \Gamma)}$$

and in this case, there is no solution such that $z_2 \neq z_1$.

Provided $d_1(z_1) \neq 0$, one gets $z_2 = r_1(z_1)$:

$$r_1(z_1) = \frac{(\varepsilon - 1)^2 \, (2\,\Gamma - \gamma)\, z_1}{d_1(z_1)}$$

which is a rational function of z_1. And the intersection with $\{D = 0\} \cap \{D' = 0\}$ is the curve $\{(0, z_1, 0, r_1(z_1))\, z_1 \in \mathbb{R} \setminus \{\tilde{z}_{\gamma, \Gamma}\}\}$.

Lemma 13 $\{D = 0\} \cap \{D' = 0\} \cap \{y_1 = 0\} \cap \{(z_1 - z_2) \neq 0\}$ *is the union of two lines of* $\{y_1 = z_1 = 0\}$ *intersecting at N and a rational curve* $\{(0, z_1, 0, r_1(z_1))\, z_1 \in \mathbb{R} \setminus \{\tilde{z}_{\gamma, \Gamma}\}\}$.

– Let us assume $y_1 \neq 0$. We can eliminate z_2 using:

$$z_2 = \frac{z_1\, y_2\, (1 - \varepsilon)}{y_1}$$

and, substituting in $y_1^2 \tilde{D}$ we get the factorization $y_1^2 \tilde{D} = q_1\, q_2$, with:

$$\begin{aligned}
q_1 = {} & \Gamma\,(\varepsilon - 1)\,(2\,\Gamma - \gamma)\, y_1^3 + \gamma^2\,(\varepsilon - 1)\, z_1^2 y_1 + \gamma^2\,(\varepsilon - 1)^2\, z_1^2 y_2 \\
& + (2\,\Gamma\,\varepsilon\,(\varepsilon - 2)\,(\gamma - \Gamma)\, z_1 - \Gamma\,(\gamma - 2\,\Gamma))\, y_2\, y_1^2
\end{aligned}$$

and:

$$\begin{aligned}
q_2 &= (\varepsilon - 1)\,(\gamma - 2\,\Gamma)\, y_1 + (2\,\varepsilon\,(2 - \varepsilon)\,(\gamma - \Gamma)\, z_1 + \gamma - 2\,\Gamma)\, y_2 \\
&= (\varepsilon - 1)\,(\gamma - 2\,\Gamma)\, y_1 + d_1(z_1) y_2.
\end{aligned}$$

Provided $d_1 \neq 0$ (that is $z_1 \neq \tilde{z}_{\gamma, \Gamma}$), we solve $q_2 = 0$ with respect to y_2, and then we get the value of (y_2, z_2):

$$\left(\frac{(\varepsilon - 1)\,(\gamma - 2\,\Gamma)\, y_1}{d_1(z_1)},\ \frac{(\varepsilon - 1)^2\,(2\,\Gamma - \gamma)\, z_1}{d_1(z_1)} \right).$$

Lemma 14 $\{D = 0\} \cap \{D' = 0\} \cap \{(z_1 - z_2)\, y_1 d_1(z_1) \neq 0\}$ *is a rational surface* $(y_2 = \rho_2(y_1, z_1), z_2 = \rho_1(z_1) y_1 \neq 0 z_1 \neq \tilde{z}_{\gamma, \Gamma})$.

We put d_3

$$d_3 = (2\,\Gamma\,\varepsilon\,(\varepsilon - 2)\,(\gamma - \Gamma)\, z_1 - \Gamma\,(\gamma - 2\,\Gamma))\, y_1^2 + \gamma^2\,(\varepsilon - 1)^2\, z_1^2$$

Its discriminant with respect to y_1 is:

$$-4\left(2\,\Gamma - 4\,\gamma\, z_1\, \varepsilon + 2\,\gamma\, z_1\, \varepsilon^2 - \gamma + 4\,\Gamma\, z_1\, \varepsilon - 2\,\Gamma\, z_1\, \varepsilon^2\right)\Gamma\, \gamma^2 z_1^2\,(\varepsilon - 1)^2$$

$$-4\,(2\,\Gamma - \gamma + 2\,\varepsilon\,(2 - \varepsilon)\,(\Gamma - \gamma)\, z_1)\,\Gamma\, \gamma^2 z_1^2\,(\varepsilon - 1)^2$$

and its sign changes when z_1 reaches $\tilde{z}_{\gamma, \Gamma}$.

Provided $d_3(y_1, z_1) \neq 0$, we solve q_1 with respect to y_2, and then we get the value of (y_2, z_2):

$$\left(\frac{\left(\Gamma \left(2\Gamma - \gamma \right) y_1^2 + \gamma^2 z_1^2 \right) (1 - \varepsilon) \ y_1}{d_3(y_1, z_1)}, \ \frac{\left(\Gamma \left(2\Gamma - \gamma \right) y_1^2 + \gamma^2 z_1^2 \right) (\varepsilon - 1)^2 \ z_1}{d_3(y_1, z_1)} \right).$$

Lemma 15 $\{D = 0\} \cap \{D' = 0\} \cap \{(z_1 - z_2) \ y_1 d_3(z_1) \neq 0\}$ *is a rational surface with parameterization* $(y_2 = \rho_3(y_1, z_1), z_2 = \rho_4(y_1, z_1))$.

- Analysis of the behaviors of the solutions of X_e^r near O.
  We set $\tilde{z}_i = 1 + z_i$ and we have the following approximations:

  - $D = (1 - \varepsilon)\tilde{D}, \ \tilde{D} = h_1 + h_2,$

  $$h_1 = \gamma^2 \varepsilon \ (\varepsilon - 2) \ (\gamma - 2\Gamma) \ (\tilde{z}_1 - \tilde{z}_2)$$
  $$h_2 = \Gamma \ (\varepsilon - 1)^2 \ (\gamma - 2\Gamma)^2 \ y_1^2 + 2\Gamma \ (\gamma - 2\Gamma)^2 \ (\varepsilon - 1) \ y_2 \ y_1$$
  $$+ \Gamma \ (\gamma - 2\Gamma)^2 \ y_2^2 - \gamma^2 \ (\varepsilon - 1)^2 \ (\gamma - 2\Gamma) \ \tilde{z}_1^2$$
  $$- \gamma^2 \ (\gamma - 2\Gamma) \ \tilde{z}_2^2 + \gamma^2 \left(\varepsilon^2 + 2 - 2\varepsilon \right) (\gamma - 2\Gamma) \ \tilde{z}_1 \tilde{z}_2.$$

  - $D' = 2\gamma^2 (\Gamma - \gamma)(2\Gamma - \gamma)(1 - \varepsilon)(\tilde{z}_2 - \tilde{z}_1)[(-1 + \tilde{z}_1) y_2 (\varepsilon - 1) + (-1 + \tilde{z}_2) y_1].$

Conclusion: these computations allow to evaluate the equilibrium points and the behaviors of the solutions near such point, using linearization methods. A first step towards the global behavior is the following result.

Lemma 16 *The surface* $y_1 = y_2 = 0$ *is foliated by lines solutions connecting O to the north pole N, the singular control being zero.*

Proposition 22 *Singular points on* $y_1 = y_2 = 0$, $z_1 = z_2 = \tilde{z}$ *are such that: in the coordinates* $\bar{q} = (y_1, y_2, z_1 - z_2, z_1)$ *the system takes the form*

$$\dot{\bar{q}} = A\bar{q} + R(\bar{q})$$

where

$$A = \begin{pmatrix} 0 & 0 & 0 & 0 \\ 0 & 0 & 0 & 0 \\ 0 & 0 & 0 & 0 \\ 0 & -\gamma^3 \varepsilon \ \bar{z}^2 \ (\varepsilon - 2) \ (2\delta + \gamma + 2\bar{z}\delta) & 0 & 0 \end{pmatrix}.$$

- *At the North Pole,* $A = 0$, $R\bar{q} = O(|\bar{q}|^3)$.
- *At the point* $S = (0, z_s, 0, z_s)$ *where* $z_s = \frac{\gamma - 2\Gamma}{2(\Gamma - \gamma)}$, $A = 0$, $R(\bar{q}) = O(|\bar{q}|^2)$.

Locally the trajectories can be computed using a blowing-up.

3.3.8 Numerical Simulations, the Ideal Contrast Problem

This section is devoted to numerical simulation in the ideal control problem using three complementary softwares:

- `Bocop`: direct method,
- `HamPath`: indirect method,
- `GloptiPoly`: Lmi technique to estimate the global optimum.

The algorithms based on the softwares are presented in details in [25].

The ideal contrast problem by saturation in the single-input case, can be summarized this way:

$$\begin{cases} c(q(t_f)) = -|q_2(t_f)|^2 \longrightarrow \min_{u(\cdot)}, \text{ fixed } t_f \\ \dot{q} \quad = F_0(q) + u_1 F_1(q), \\ q(0) \quad = q_0 \\ q_1(t_f) \quad = 0 \end{cases} \quad \text{(ICPS)}$$

where $q = (q_1, q_2)$, $q_i = (y_i, z_i) \in \mathbb{R}^2$, $|q_i| \leq 1$, $i = 1, 2$. The initial condition for each spin is $q_i(0) = (0, 1)$. The vector fields F_0 and F_1 are given by:

$$F_0(q) = \sum_{i=1,2}(-\Gamma_i y_i)\frac{\partial}{\partial y_i} + (\gamma_i(1 - z_i))\frac{\partial}{\partial z_i},$$

$$F_1(q) = \sum_{i=1,2}-z_i\frac{\partial}{\partial y_i} + y_i\frac{\partial}{\partial z_i},$$

where $\Lambda_i = (\gamma_i, \Gamma_i)$ are the physical parameters representing each spin.

We present the simulations using the numerical methods (see [25] for a complete description of the algorithms).

The simulations correspond to the two following sets of experimental data, with the relaxation times in seconds and T_{min} the solution of the time minimal saturation problem for a single spin, from Sect. 3.3.3.

P_1: **Fluid case**.
    Spin 1: Cerebrospinal fluid: $T_1 = 2$, $T_2 = 0.3$;
    Spin 2: Water: $T_1 = 2.5 = T_2$. $T_{min} = 26.17040$.
P_2: **Blood case**.
    Spin 1: Deoxygenated blood: $T_1 = 1.35$, $T_2 = 0.05$;
    Spin 2: Oxygenated blood: $T_1 = 1.35$, $T_2 = 0.2$.
    $T_{min} = 6.7981$.

Optimal solutions of the contrast problem are concatenations of bang and singular extremals. For the following sections, we introduce some notations. We note BS the sequence composed by one bang arc (δ_+ or δ_-) followed by one singular arc (δ_s), and nBS, $n > 1$, the concatenation of n BS-sequences.

First Results with Fixed Final Time

The first difficulty comes from the discontinuities of the optimal control structure. We need to know the control structure (meaning the number of Bang-Singular sequences) before calling the multiple shooting method. The indirect method also typically requires a reasonable estimate for the control switching times, as well as the states and costates values at the initial and switching times. We use the `Bocop` software based upon direct methods to obtain approximate optimal solutions in order to initialize the indirect shooting, within the `HamPath` code. We recall that the costate (or adjoint state) for Pontryagin's Principle corresponds to the Lagrange multipliers for the dynamics constraints in the discretized problem, and can therefore be extracted from the solution of the direct method.

The only a priori information is the value of the minimum time transfer T_{min}, used to set the final time t_f in the $[T_{min}, 3T_{min}]$ range. We note $t_f = \lambda T_{min}$ with λ in $[1, 3]$. The state variables are initialized as constant functions equal to the initial state, i.e. $y_1(\cdot) = y_2(\cdot) = 0$, $z_1(\cdot) = z_2(\cdot) = 1$. For the control variables we use the three constant initializations $u_1(\cdot) \in \{0.1, 0.25, 0.5\}$. The discretization method used is implicit midpoint (2nd order) with a number of time steps set to $\lambda \times 100$. In order to improve convergence, we add a small regularization term to the objective to be minimized, $\varepsilon_{reg} \int_0^{t_f} |u(t)|^2 dt$, with $\varepsilon_{reg} = 10^{-3}$.

We repeat the optimizations for λ in $\{1.1, 1.5, 1.8, 2.0, 3.0\}$ with the three control initializations, see Table 3.1. The solutions from `Bocop` are used to initialize the continuations in `HamPath`, and we discuss in the following sections the results obtained with the indirect method. Both methods confirm the existence of many local solutions, as illustrated on Fig. 3.2 for $\lambda = 1.5$, due in particular to symmetry reasons.

Second Order Conditions

According to proposition 3.2 from [26], the non-existence of conjugate points on each singular arc of a candidate solution is a necessary condition of local optimality. See [26] for details about conjugate points in the contrast problem. Here, we compute for each singular arc of all the solutions from Sect. 3.3.8, the first conjugate point along the arc, applying the algorithm presented in Sect. 4.3 from [26]. None of the

Table 3.1 Fluid case: Batch optimizations (Direct method). For each value of λ we test the three initializations for the control u, and record the value of the objective (i.e. the contrast), as well as the control structure (i.e. the signs of bang arcs). CPU times for a single optimization are less than one minute on a Intel Xeon 3.2 GHz

λ	1.1	1.5	1.8	2	3
u_{init} : 0.1	0.636 (++)	0.678 (+ − +)	0.688 (+ − +)	0.702 (−+)	0.683 (− + −+)
u_{init} : 0.25	FAIL	0.661 (+ + −+)	0.673 (+ + −+)	0.691 (− + +)	0.694 (+ − +)
u_{init} : 0.5	0.636 (++)	0.684 (++)	0.699 (−+)	0.697 (++)	0.698 (++)

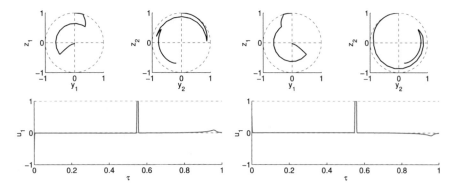

Fig. 3.2 Fluid case: Two local solutions for $\lambda = 2.0$. Trajectories for spin 1 and 2 in the (y, z)-plane are portrayed in the first two subgraphs of each subplot. The corresponding control is drawn in the bottom subgraph. The two bang arcs have the same sign for the left solution, whereas for the right solution, the two bang arcs are of opposite sign

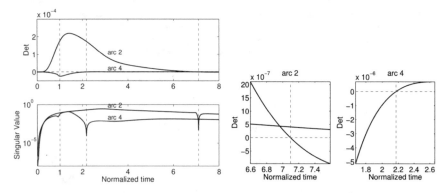

Fig. 3.3 Fluid case: second order conditions. Second order necessary condition checked on the best solution with $\lambda = 2.0$ from Sect. 3.3.8. The rank condition from the algorithm presented in Sect. 4.3 from [26] is evaluated along the two singular arcs. See [21] for details on the concept of conjugate times. On the left subplot, for each singular arc, the curve is reparameterized so that the final time corresponds to the abscissa 1 (vertical blue dashed line); the determinant associated with the rank condition is plotted (top subgraph), so there is a conjugate time whenever it vanishes (vertical red dashed lines). One observes that conjugate times on each arc are located after the (normalized to 1) final time, satisfying necessary condition of local optimality of the trajectory. At the bottom, the smallest singular value of the matrix whose rank we test is plotted, extracting only the relevant information to detect the rank drops. On the right subplot is presented a zoom of top-left subgraph near the two conjugate times

solutions has a conjugate point on a singular arc. Hence all the solutions satisfy the second order necessary condition of local optimality. Figure 3.3 represents the computations of the two conjugate points (since the structure is 2BS) of the best solution with $\lambda = 2.0$ from Sect. 3.3.8.

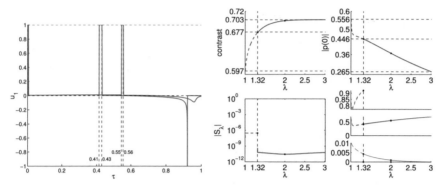

Fig. 3.4 Fluid case: influence of the final time. On the left subgraph are shown the control laws of solutions at $\lambda = 2$ and $\lambda = 1.32$ from path from the right subplot. For $\lambda = 1.32$, we can see the saturating singular arc around the normalized time $\tau = 0.92$ (the time is normalized to be between 0 and 1 for each solution). The 2BS solution at $\lambda = 1.32$ is used to initialize a multiple shooting with a 3BS structure and then to perform a new homotopy from $\lambda = 1.32$ to $\lambda = 1$. On the right subgraph is portrayed the two homotopies: the first from $\lambda = 2$ to $\lambda = 1.32$ and the second to $\lambda = 1$, with one more BS sequence. The value function, the norm of the initial adjoint vector, the norm of the shooting function and the switching times along the path are given. The blue color represents 2BS solutions while the red color is for 3BS structures. The dashed red lines come from the extended path after the change of structure detected around $\lambda = 1.32$

Influence of the Final Time

Given that the initial point (the North pole) is a stationary point, the constrast is an increasing function of t_f acting as a parameter. Indeed, applying a zero control at $t = 0$ leaves the system in its initial state so there is an inclusion of admissible controls between problems when the final time is increased (and the bigger the set of controls, the larger the maximum contrast). Having increasing bounded (by one, which is the maximum possible contrast given the final condition on spin no. 1) functions, it is natural to expect asymptotes on each branch.

In both cases P_1 and P_2, the contrast problem has many local solutions, possibly with different control structures. Besides, the structure of the best policy can change depending on the final time. The possible change of structure along a single path of zeros is emphasized in Fig. 3.4. In this figure, the branch made of 2BS solutions is represented in blue, whereas the 3BS branch is the dashed red line. We also show a crossing between two value functions of two different paths of zeros in Fig. 3.5.

Then for each solution of each branch the second order necessary condition is checked as in Sect. 3.3.8: the first conjugate point of each singular extremal is computed. There is no failure in this test, hence all the solutions satisfy the necessary second order condition of local optimality. Figure 3.6 presents the second order conditions along the extended path from Fig. 3.4.

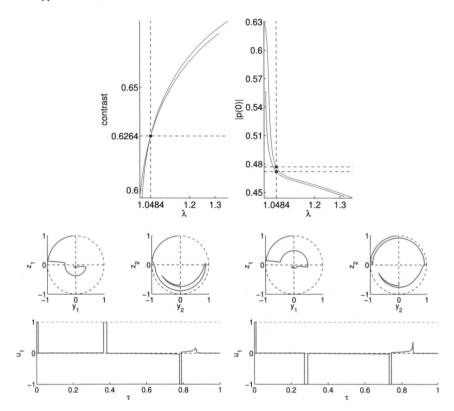

Fig. 3.5 Fluid case: influence of the final time. Crossing between two branches with 3BS solutions. The crossing is around $\lambda = 1.0484$, see top subgraph. Thus for $\lambda \leq 1.0484$, the best solution, locally, has a 3BS structure of the form $\delta_+\delta_s\delta_+\delta_s\delta_-\delta_s$ (bottom-left subgraph) while for $\lambda \in [1.0484, 1.351]$ the best solution is of the form $\delta_+\delta_s\delta_-\delta_s\delta_-\delta_s$ (bottom-right subgraph). On the two bottom subgraphs, the trajectories for spin 1 and 2 in the (y, z)-plane are portrayed with the corresponding control, both for $\lambda = 1.0484$

Sub-optimal Syntheses in Fluid and Blood Cases

We give the syntheses of locally optimal solutions obtained in the blood and fluid cases. Note that in the special case $t_f = T_{\min}$, for both cases the solution is 2BS and of the form $\delta_+\delta_s\delta_+\delta_s$.

For the fluid case, the left subplot of Fig. 3.7 represents all the different branches we obtained by homotopy on λ. The greatest two value functions intersect around $t_f = 1.048 T_{\min}$. The right subplot shows the sub-optimal synthesis. The best policy is:

$$
\begin{aligned}
\delta_+\delta_s\delta_+\delta_s &\quad \text{for } \lambda \in [1.000, 1.006], \\
\delta_+\delta_s\delta_+\delta_s\delta_-\delta_s &\quad \text{for } \lambda \in [1.006, 1.048], \\
\delta_+\delta_s\delta_-\delta_s\delta_-\delta_s &\quad \text{for } \lambda \in [1.048, 1.351], \\
\delta_+\delta_s\delta_-\delta_s &\quad \text{for } \lambda \in [1.351, 3.000].
\end{aligned}
\tag{3.37}
$$

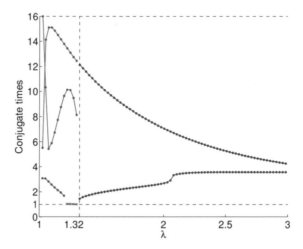

Fig. 3.6 Fluid case: influence of the final time. Second order necessary condition checked along the extended path from Fig. 3.4. For all solutions from $\lambda = 1$ to $\lambda = 3$ are computed the first conjugate times along each singular arc. For $\lambda \in [1, 1.32]$, the structure is 3BS and there are 3 singular arcs. For $\lambda \in [1.32, 3]$, there are 2 singular arcs. Each singular interval is normalized in such a way the initial time is 0 and the final time is 1. The lower dashed horizontal line represents the final time 1. There is no conjugate time before the normalized final time 1 which means that all solutions satisfy the second order necessary condition of local optimality. Note that at a magenta cross, around $(1.32, 1)$, the control of the first singular arc saturates the constraint $|u| = 1$, and so no conjugate time is computed after this time

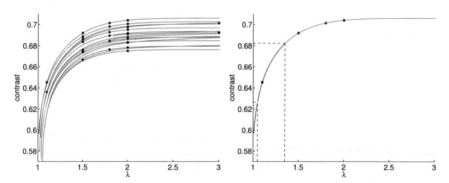

Fig. 3.7 Fluid case, sub-optimal synthesis. Illustration on the left subplot, of local solutions (each branch corresponds to a control structure). The suboptimal synthesis is plotted on right subplot. The colors are blue for 2BS structure, red for 3BS and green for 4BS. The best policy is $\delta_{+}\delta_{s}\delta_{+}\delta_{s}\delta_{-}\delta_{s}$ for $\lambda \leq 1.0484$, and $\delta_{+}\delta_{s}\delta_{-}\delta_{s}\delta_{-}\delta_{s}$ for $\lambda \in [1.0484, 1.351]$. Then, for $\lambda \in [1.351, 3]$, the best policy is 2BS and of the form $\delta_{+}\delta_{s}\delta_{-}\delta_{s}$

For the blood case, the results are excerpted from [38]. The left subplot of Fig. 3.8 shows the contrast for five different components of $\{h = 0\}$, for final times $t_f \in [1, 2]T_{\min}$. The three black branches are made only of BS solutions whereas the two

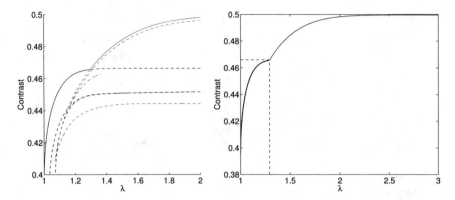

Fig. 3.8 Blood case, sub-optimal synthesis. Illustration on the left subplot, of local solutions (each branch corresponds to a control structure).Best policy as solid lines, local solutions as dashed lines. The suboptimal synthesis is plotted on right subplot. The colors are black for BS structure, blue for 2BS and red for 3BS. The best policy is BS for $t_f \in (1, 1.294)T_{\min}$ and 3BS of the form $\delta_+ \delta_s \delta_- \delta_s \delta_- \delta_s$ for $t_f \in (1.294, 2]T_{\min}$. In the special case $t_f = T_{\min}$, the solution is 2BS and of the form $\delta_+ \delta_s \delta_+ \delta_s$

others are made of 2BS and 3BS solutions. To maximize the contrast, the best policy, drawn as solid lines, is:

$$
\begin{aligned}
&\delta_+\delta_s\delta_+\delta_s && \text{for } \lambda \in [1.000, 1+\varepsilon], \ \varepsilon > 0 \text{ small} \\
&\delta_+\delta_s && \text{for } \lambda \in [1+\varepsilon, 1.294], && (3.38) \\
&\delta_+\delta_s\delta_-\delta_s\delta_-\delta_s && \text{for } \lambda \in [1.294, 2.000].
\end{aligned}
$$

Sub-optimal Syntheses Compared to Global Results
We now apply the `lmi` method to the contrast problem, described in [25], in order to obtain upper bounds on the true contrast. Comparing these bounds to the contrast of our solutions then gives an insight about their global optimality.

Table 3.2 shows the evolution of the upper bound on the contrast in function of `lmi` relaxation order,for the fluid case with $t_f = T_{\min}$. As expected, the method yields a monotonically non-increasing sequence of sharper bounds. Relaxations of orders 4 and 5 yield very similar bounds, but this should not be interpreted as a termination criterion for the `lmi` method.

Figures 3.9 and 3.10 compare the tightest upper bounds found by the `lmi` method with the best candidate solutions found by `Bocop` and `HamPath`, in both the blood and fluid cases. The figures also represent the relative gap between the methods defined as $(C_{LMI} - C_H)/C_H$, where C_{LMI} is the `lmi` upper bound and C_H is the contrast found with `HamPath`. As such, this measure characterizes the optimality gap between the methods. It does not, however, specify which of the method(s) could be further improved. At the fifth relaxation, the average gap is around 11% in the

Table 3.2 Fluid case, $t_f = T_{min}$: upper bounds on contrast $\sqrt{-J_M^r}$, numbers of moments N_r and CPU times t_r in function of relaxation order r

r	$\sqrt{-J_M^r}$	N_r	t_r
1	0.8474	63	0.7
2	0.7552	378	3
3	0.6226	1386	14
4	0.6069	3861	332
5	0.6040	9009	8400

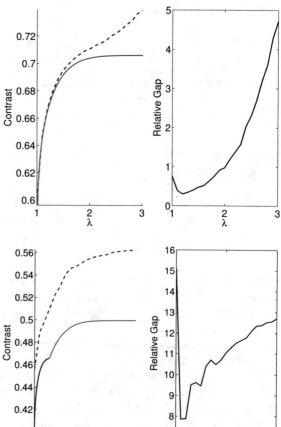

Fig. 3.9 Fluid case. Best upper bounds (dashed line) by the `lmi` method compared with best solutions by HamPath (solid line), and relative gap between the two

Fig. 3.10 Blood case. Best upper bounds (dashed line) by the `lmi` method compared with best solutions by HamPath (solid line), and relative gap between the two

blood case, which, given the application, is satisfactory on the experimental level. For the fluid case, the average gap on the contrast is about 2% at the fifth relaxation, which strongly suggest that the solution is actually a global optimum. The gap is even below the 1% mark for $t_f \leq 2\,T_{min}$.

3.3.9   Numerical Simulations, the Multisaturation of Two Spins with B_1-inhomogeneity

In this section we give an illustration of our techniques applied to the saturation of two spins combining geometric analysis and numerical simulations to deduce the solution. We proceed in two steps.

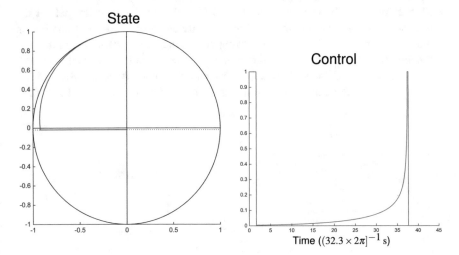

Fig. 3.11 Time-minimal saturation of a single spin ($\lambda = 0$)

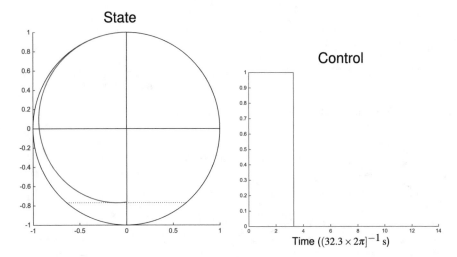

Fig. 3.12 Time-minimal saturation of a single spin ($\lambda = 0.9941 \simeq \bar{\lambda}$)

- **Step 1: Time minimal saturation of a single spin** In the single-spin case the time minimal solution is described in Fig. 3.1 leads to construct the optimal solution for a continuation on the set of parameters where $\lambda = 0$ corresponds to the case of deoxygenated blood, $\lambda = 1$ corresponds to the case $2\Gamma = 3\gamma$ and $\lambda = \lambda_f$ is the water case: $\Gamma = \gamma$. According to Figs. 3.11, 3.12 and 3.13, due to the control bound, the bifurcation occurs not exactly at $\lambda = 1$ when the horizontal singular line $z = \gamma/2/\delta$ leaves the Bloch ball but at $\bar{\lambda} \simeq 0.99$, since for $\lambda > \bar{\lambda}$ this line is no more accessible from the north pole.

- **Step 2**: We describe in Figs. 3.14, 3.15, 3.16 and 3.17 the BC-extremal for the multisaturation problem with B_1-inhomogeneity using the same continuation on the set of parameters. The control is computed using `HamPath` software in combination with `Bocop` in order to determine the structure of the extremal trajectory for $\lambda = 0$. Figures 3.14, 3.15 show a control with the same structure $\delta_+\delta_s\delta_+\delta_s\delta_+\delta_s$, that is a sequence of three bang-singular arcs. A bifurcation occurs at $\bar{\lambda} \simeq 0.94$

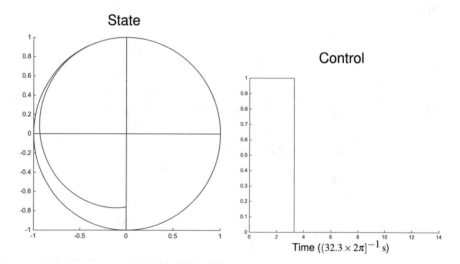

Fig. 3.13 Time-minimal saturation of a single spin ($\lambda = \lambda_f$)

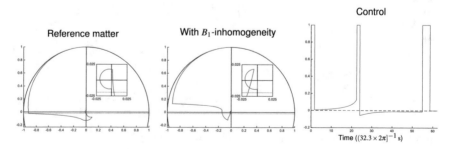

Fig. 3.14 BC-extremal for the multisaturation problem with $\lambda = 0$

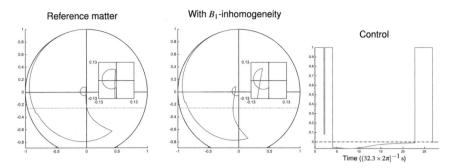

Fig. 3.15 BC-extremal for the multisaturation problem with $\lambda = 0.943 < \bar{\lambda}$

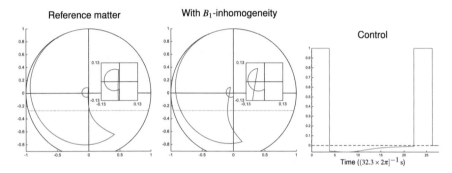

Fig. 3.16 BC-extremal for the multisaturation problem with $\lambda = 0.948 > \bar{\lambda}$

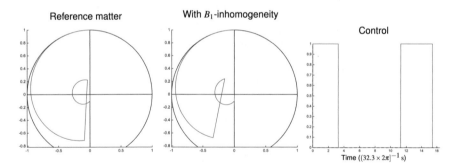

Fig. 3.17 BC-extremal for the multisaturation problem with $\lambda = \lambda_f$

where the first singular arc disappears. Figures 3.16, 3.17 show a control with a different structure $\delta_+\delta_s\delta_+\delta_s$. In each picture, we have represented the critical altitude $z = \gamma/2/\delta$ (on horizontal dotted line). At $\lambda = \lambda_f$, the extremal is simply $\delta_+\delta_0\delta_+\delta_0$: singular arcs are obtained by applying a zero control.

Chapter 4
Conclusion

The two cases studied in this book show the practical interest of combining geometric optimal control with numeric computations using the developed software to solve industrial type problems.

The application to microswimmers is very recent and validate results obtained from fluid mechanics practitioners based on curvature control and Fourier analysis. The SR-geometry framework allows to compare different strokes and *different swimmers*, using the mechanical energy cost. The copepod mathematical swimmer is the simplest slender body model. Normal and abnormal strokes have interpretation in terms of sinusoidal and sequential paddlings. This leads to design a simple macroscopic copepod robot to validate the theoretical computations of the most efficient stroke. Another validation of the mathematical model using Resistive Force Theory for Stokes' flow is coming from the observations [65] showing the agreement between observed and predicted displacements. The mathematical developments lead to solve the inverse problem of identifying the cost used for the copepod nauplii displacement.

The developments motivated by MRI are more profound and lead to intricate numerical investigations to deal with an highly complex optimal control problem with many local optimal solutions. Nevertheless we believe that the techniques validate by in vitro and in vivo experiments realized under the auspices of the ANR project DFG Explosy will find in a very near future applications in MRI diagnosis.

References

1. Agrachev, A., Chtcherbakova, N.N., Zelenko, I.: On curvatures and focal points of dynamical Lagrangian distributions and their reductions by first integrals. J. Dyn. Control Syst. **11**(3), 297–327 (2005)
2. Agrachev, A., Gauthier, J.P.: On the Dido problem and plane isoperimetric problems. Acta Appl. Math. **57**(3), 287–338 (1999)
3. Agrachev, A., Sarychev, A.: Abnormal sub-Riemannian geodesics: morse index and rigidity. Ann. Inst. H. Poincaré Anal. Non Linéaire **13**(6), 635–690 (1996)
4. Aleexev, V., Tikhomirov, V., Fomine, S.: Commande Optimale. Mir, Moscow (1982)
5. Alouges, F., DeSimone, A., Giraldi, L., Zoppello, M.: Self-propulsion of slender micro-swimmers by curvature control: N-link swimmers. Int. J. of Non-Linear Mech. **56**, 132–141 (2013)
6. Alouges, F., DeSimone, A., Lefebvre, A.: Optimal strokes for low Reynolds number swimmers: an example. J. Nonlinear Sci. **18**, 277–302 (2008)
7. Arcostanzo, M., Arnaud, M.-C., Bolle, P., Zavidovique, M.: Tonelli Hamiltonians without conjugate points and C^0-integrability. Math. Z. **280**(1–2), 165–194 (2015)
8. Arnold, V. I., Guseĭn-Zade, S.M., Varchenko, A.N.: Singularities of Differentiable Maps. Birkhäuser Boston, Inc., Boston MA (1985)
9. Arnold, V.I.: Mathematical methods of Classical Mechanics. 2 edn. Graduate Texts in Mathematics, 60, p 508. Springer-Verlag, New York (1989)
10. Avron, J.E., Raz, O.: A geometric theory of swimming: Purcell's swimmer and its symmetrized cousin. New J. Phys. **10**(6), 063016 (2008)
11. Batchelor, G.K.: Slender-body theory for particles of arbitrary cross-section in Stokes flow. J. Fluid Mech. **44**, 419–440 (1970)
12. Becker, L.E., Koehler, S.A., Stone, H.A.: On self-propulsion of micro-machines at low Reynolds number: Purcell's three-link swimmer. J. Fluid Mech. **490**, 15–35 (2003)
13. Bellaïche, A.: The tangent space in sub-Riemannian geometry. J. Math. Sci. (New York) **35**, 461–476 (1997)
14. Berger, M.: La taxonomie des courbes. Pour la Sci. **297**, 56–63 (2002)
15. Bettiol, P., Bonnard, B., Giraldi, L., Martinon, P., Rouot, J.: The three links Purcell swimmer and some geometric problems related to periodic optimal controls. Rad. Ser. Comp. App. **18** (2016). (Variational Methods, Ed. by M. Bergounioux et al.)
16. Bettiol, P., Bonnard, B., Nolot, A., Rouot, J.: Sub-Riemannian Geometry and Swimming at Low Reynolds Number: The Copepod Case. Accepted in ESAIM Control Optim, Calc (2017)

B. Bonnard et al., *Geometric and Numerical Optimal Control*, SpringerBriefs in Mathematics, https://doi.org/10.1007/978-3-319-94791-4

17. Bettiol, P., Bonnard, B., Rouot, J.: Optimal strokes at low Reynolds number: a geometric and numerical study of Copepod and Purcell swimmers, to appear SICON (2018)
18. Bloch, F.: Nuclear induction. Phys. Rev. **7–8**, 460 (1946)
19. Bonnans, F., Giorgi, D., Maindrault, S., Martinon, P., Grélard, V.: Bocop—A collection of examples, Inria Research Report, Project-Team Commands. **8053** (2014)
20. Bonnard, B.: Feedback equivalence for nonlinear systems and the time optimal control problem. SIAM J. Control Optim. **29**(6), 1300–1321 (1991)
21. Bonnard, B., Caillau, J.-B., Trélat, E.: Second order optimality conditions in the smooth case and applications in optimal control. ESAIM Control Optim. Calc. Var. **13**(2), 207–236 (2007)
22. Bonnard, B., Chyba, M.: Singular Trajectories and their Role in Control Theory. Springer-Verlag, Berlin (2003)
23. Bonnard, B., Chyba, M., Jacquemard, A., Marriott, J.: Algebraic geometric classification of the singular flow in the contrast imaging problem in nuclear magnetic resonance. Math. Control Relat. Fields **3**(4), 397–432 (2013)
24. Bonnard, B., Chyba, M., Marriott, J.: Singular trajectories and the contrast imaging problem in nuclear magnetic resonance. SIAM J. Control Optim. **51**(2), 1325–1349 (2013)
25. Bonnard, B., Claeys, M., Cots, O., Martinon, P.: Geometric and numerical methods in the contrast imaging problem in nuclear magnetic resonance. Acta Appl, Math (2013)
26. Bonnard, B., Cots, O.: Geometric numerical methods and results in the control imaging problem in nuclear magnetic resonance. Math. Models Methods Appl. Sci. **24**(1), 187–212 (2014)
27. Bonnard, B., Faubourg, L., Trélat, E.: Mécanique céleste et contrôle des véhicules spatiaux. Mathématiques & Applications, Springer-Verlag **51**, Berlin (2006)
28. Bonnard, B., Jacquemard, A., Rouot, J.: Optimal control of an ensemble of bloch equations with applications in MRI. In: 2016 IEEE 55th Conference on Decision and Control (CDC), Las Vegas, NV, USA (2017)
29. Bonnard, B., Kupka, I.: Théorie des singularités de l'application entrée/sortie et optimalité des trajectoires singulières dans le problème du temps minimal. Forum Math. **5**(2), 111–159 (1993)
30. Bliss, G.A.: Lectures on the Calculus of Variations. University of Chicago Press, Chicago (1946)
31. Brockett, R.W.: Control theory and Singular Riemannian Geometry. Springer, New York-Berlin, pp. 11–27 (1982)
32. Caillau, J.-B., Cots, O., Gergaud, J.: Differential continuation for regular optimal control problems. Optim. Methods Softw. **27**(2), 177–196 (2012)
33. Cartan, E.: Les systèmes de Pfaff a cinq variables et les équations aux derivées partielles du second ordre. Ann. Sci. École Normale **27**, 109–192 (1910)
34. Chambrion, T., Giraldi, L., Munnier, A.: Optimal strokes for driftless swimmers: a general geometric approach. Accepted in ESAIM Control Optim. Calc, Var (2017)
35. Chitour, Y., Jean, F., Trélat, E.: Genericity results for singular curves. J. Diff. Geom. **73**(1), 45–73 (2006)
36. Chow, W.L.: Über Systeme von linearen partiellen Differentialgleichungen erster Ordnung. Math. Ann. **117**, 98–105 (1939)
37. Conolly, S., Nishimura, D., Albert, A.: Optimal control solutions to the magnetic resonance selective excitation problem. Med. Imaging IEEE Trans. **5**(2), 106–115 (1986)
38. Cots, O.: Contrôle optimal géométrique: méthodes homotopiques et applications. PhD thesis, Université de Bourgogne (2012)
39. Dieudonné, J.A., Carrell, J.B.: Invariant Theory, Old and New. Academic Press, New York, London (1971)
40. Gamkrelidze, R.V.: Discovery of the maximum principle. J. Dynam. Control Syst. **5**(4), 437–451 (1977)
41. Gelfand, I.M., Fomin, S.V.: Calculus Var. Prentice Hall Inc., Englewood Cliffs, New Jersey (1963)
42. Godbillon, C.: Geométrie différentielle et mécanique analytique. Hermann, Paris (1969)
43. Gregory, J.: Quadratic form theory and differential equations. Math. Sci. Eng. **152**, New York, London (1980)

44. Hancock, G.J.: The self-propulsion of microscopic organisms through liquids. Proc. R. Soc. Lond. A **217**, 96–121 (1953)
45. Happel, J., Brenner, H.: Low Reynolds Number Hydrodynamics with Special Applications to Particulate Media. Prentice-Hall Inc., Englewood Cliffs, N.J. (1965)
46. Helgason, S.: Differential Geometry, Lie Groups, and Symmetric Spaces, Pure and Applied Mathematics, 80, p. 628. Academic Press Inc., New York, London (1978)
47. Henrion, D., Lasserre, J.-B.: GloptiPoly: global optimization over polynomials with Matlab and SeDuMi. ACM Trans. Math. Softw. **29**(2), 165–194 (2003)
48. Hermes, H.: Lie algebras of vector fields and local approximation of attainable sets. SIAM J Control Optim. **16**(5), 715–727 (1978)
49. Jean, F.: Control of Nonholonomic Systems: from Sub-Riemannian Geometry to Motion Planning. Springer International Publishing, Springer Briefs in Mathematics (2014)
50. John, F.: Partial differential equations, reprint of 4th edition. Appl. Math. Sci. **1**, Springer-Verlag, New York (1991)
51. Jurdjevic, V.: Geometric control theory. Camb. Stud. Adv. Math. **52**, Cambridge University Press, Cambridge (1997)
52. Klingenberg, W.: Riemannian Geometry. de Gruyter Studies in Mathematics. Walter de Gruyter and Co., Berlin, New York (1982)
53. Krener, A.J.: The high order maximum principle and its application to singular extremals. SIAM J. Control Optim. **15**(2), 256–293 (1977)
54. Kupka, I.: Geometric theory of extremals in optimal control problems. i. the fold and Maxwell case. Trans. Amer. Math. Soc. **299**(1), 225–243 (1987)
55. Kupka, I.: Géométrie sous-riemannienne. Astérisque, Séminaire Bourbaki **1995**(96), 351–380 (1997)
56. Landau, L., Lipschitz, E.: Physique théorique. Ed. Mir (1975)
57. Lapert, M.: Développement de nouvelles techniques de contrôle optimal en dynamique quantique: de la Résonance Magnétique Nucléaire à la physique moléculaire. Phd thesis, Laboratoire Interdisciplinaire Carnot de Bourgogne, Dijon (2011)
58. Lapert, M., Zhang, Y., Glaser, S.J., Sugny, D.: Towards the time-optimal control of dissipative spin-1/2 particles in nuclear magnetic resonance. J. Phys. B: At. Mol. Opt. Phys. **44**, 15 (2011)
59. Lapert, M., Zhang, Y., Janich, M.A., Glaser, S.J., Sugny, D.: Exploring the physical limits of saturation contrast in magnetic resonance imaging. Sci. Reports **2** (2012)
60. Lasserre, J.-B.: Moments, Positive Polynomials and their Applications. Imperial College Press Optimization Series, Imperial College Press, London **1**, xxii+361 (2010)
61. Lasserre, J.-B., Henrion, D., Prieur, C., Trélat, E.: Nonlinear optimal control via occupation measures and LMI-relaxations. SIAM J. Control Optim. **47**(4), 1643–1666 (2008)
62. Lauga, E., Powers, T.R.: The hydrodynamics of swimming microorganisms. Rep. Progr. Phys. **72**, 9 (2009)
63. Lawden, D.F.: Elliptic functions and applications. Appl. Math. Sci. **80** (1989). (Springer-Verlag, New York)
64. Lee, E.B., Markus, L.: Foundations of Optimal Control Theory, 2nd edn. Robert E. Kreiger Publishing Co., Inc, Melbourne (1986)
65. Lenz, P.H., Takagi, D., Hartline, D.K.: Choreographed swimming of copepod nauplii. J. Royal Soc. Interface **12**, 112 20150776 (2015)
66. Levitt, M.H.: Spin Dynamics: Basics of Nuclear Magnetic Resonance. Wiley (2001)
67. Li, J.-S., Khaneja, N.: Ensemble control of Bloch equations. IEEE Trans. Automat. Control **54**(3), 528–536 (2009)
68. Liberzon, D.: Calculus of Variations and Optimal Control Theory: A Concise Introduction. Princeton University Press, (2011)
69. Lighthill, M.J.: Note on the swimming of slender fish. J. Fluid Mech. **9**, 305–317 (1960)
70. Lohéac, J., Scheid, J.-F., Tucsnak, M.: Controllability and time optimal control for low Reynolds numbers swimmers. Acta Appl. Math. **123**, 175–200 (2013)
71. Maciejewski, A.J., Respondek, W.: The nilpotent tangent 3-dimensional sub-Riemannian problem is nonintegrable. In: 2004 43rd IEEE Conference on Decision and Control (2004)

72. Milnor, J.: Morse theory. Ann. Math. Stud. **51** (1963). (Princeton University Press, Princeton)
73. Mishchenko, A.S., Shatalov, V.E., Sternin, B.Y.: Lagrangian Manifolds and the Maslov Operator. Springer Series in Soviet Mathematics. Springer-Verlag, Berlin (1990)
74. Montgomery, R.: Isoholonomic problems and some applications. Commun. Math. Phys. **128**(3), 565–592 (1990)
75. Or, Y., Zhang, S., Murray, R.M.: Dynamics and stability of low-Reynolds-number swimming near a wall. SIAM J. Appl. Dyn. Syst. **10**, 1013–1041 (2011)
76. Passov, E., Or, Y.: Supplementary notes to: Dynamics of Purcells three-link microswimmer with a passive elastic tail. EPJ E **35**, 1–9 (2012)
77. Pontryagin, L.S., Boltyanskii, V.G., Gamkrelidze, R.V.: The Mathematical Theory of Optimal Processes. John Wiley and Sons, New York (1962)
78. Purcell, E.M.: Life at low Reynolds number. Am. J. Phys. **45**, 3–11 (1977)
79. Rashevskii, P.K.: About connecting two points of complete non-holonomic space by admissible curve. Uch. Zapiski ped. inst. Libknexta **2**, 83–94 (1938)
80. Rouot, J., Bettiol, P., Bonnard, B., Nolot, A.: Optimal control theory and the efficiency of the swimming mechanism of the Copepod Zooplankton. To appear in Proceedings 20th IFAC World Congress, Toulouse (2017)
81. Sachkov, Y.L.: Symmetries of flat rank two distributions and sub-Riemannian structures. Trans. Amer. Math. Soc. **356**, 457–494 (2004)
82. Schättler, H., Ledzewicz, U.: Geometric optimal control. Theory, methods and examples. Interdisciplinary Applied Mathematics, 38. Springer, New York (2012)
83. Skinner, T.E., Reiss, T., Luy, B., Khaneja, N., Glaser, S.J.: Application of optimal control theory to the design of broadband excitation pulses for high-resolution NMR. J. Magn. Reson. **163**(1), 8–15 (2003)
84. Sontag, E.D.: Mathematical control theory. Deterministic finite-dimensional systems, second edition. Texts in Applied Mathematics **6**, Springer-Verlag, New York (1998)
85. Sussmann, H.J.: Orbits of families of vector fields and integrability of distributions. Trans. Am. Math. Soc. **180**, 171–188 (1973)
86. Sussmann, H.J., Jurdjevic, V.: Controllability of non-linear systems. J Differential Equ. **12**, 95–116 (1972)
87. Takagi, D.: Swimming with stiff legs at low Reynolds number. Phys. Rev. E **92** (2015)
88. Trélat, E.: Contrôle optimal, Mathématiques Concrètes. [Concrete Mathematics], pp. vi+246 (2005)
89. Vinter, R.: Optimal control. Systems & Control: Foundations & Applications, Birkhäuser Boston, Inc., Boston, MA xviii+507 (2000)
90. Zakaljukin, V.M.: Lagrangian and Legendre singularities. Funkcional. Anal. i Priložen. **10**, 26–36 (1976)
91. Zhitomirskiĭ, M.: Typical singularities of differential 1-forms and Pfaffian equations. Am. Math. Soc. Providence, RI. **113**, 176 (1992)
92. Zhu, J., Trélat, E., Cerf, M.: Geometric optimal control and applications to aerospace, Pacific J. Math. for Industry, **9**, 8 (2017)

Printed in the United States
By Bookmasters